To Pamela

The Birth of British Aviation

"Prisoners of Hope"

HELEN LANDAU

With Very Best Wishes
and Thanks,

Helen L

First published in Great Britain in August 2013 by

MG Powell Company,
6, Evelyn Court,
Malvern Road,
Cheltenham GL50 2JR
email: helen.landau@tinyworld.co.uk
www.helenlandau.co.uk

ISBN: 978-0-9926506-0-5

Cover design by Alan Hunns
Layout design by Marie Doherty
Copy/editing by Ian Howe

CONTENTS

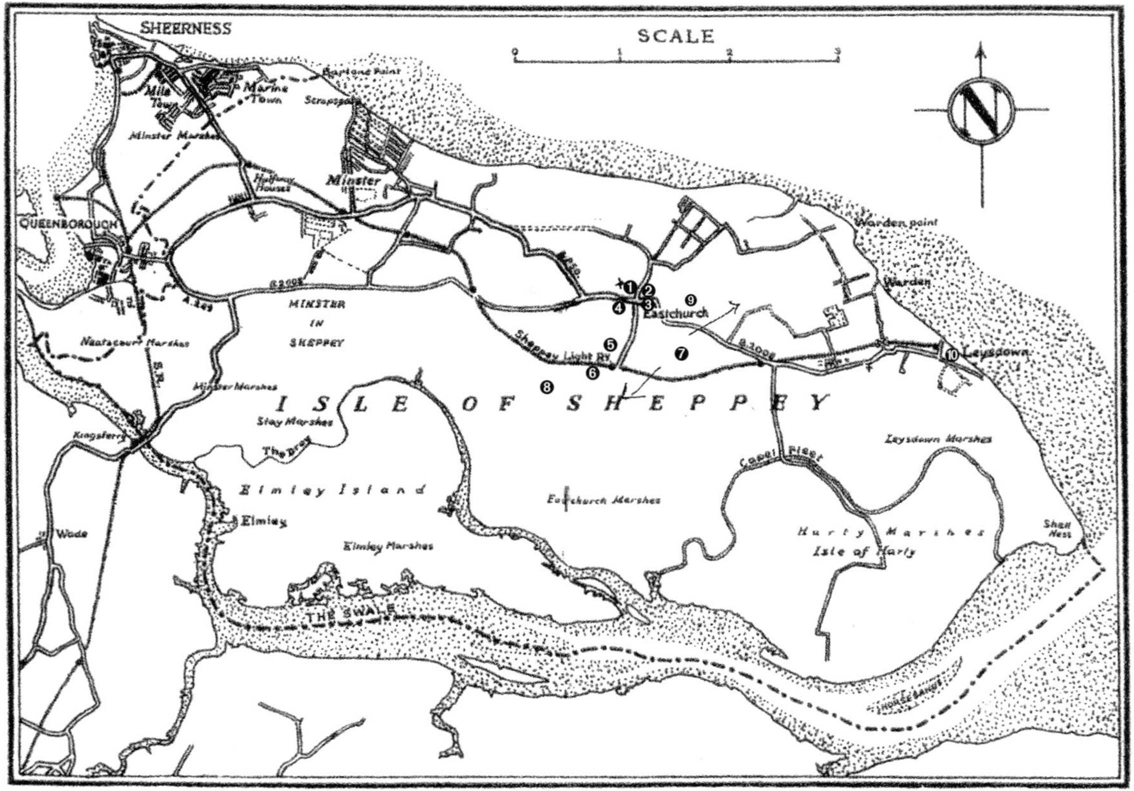

❶ All Saints Church

❷ The Castle

❸ The Billet

❹ Pioneer Memorial

❺ Parsonage Farm

❻ Standford Hill

❼ Stone Pitts Flying Fields

❽ Flying Sheds

❾ Shurland Hall

❿ Mussel Manor

Map of locations

CHAPTER ONE

Charlie Rolls

The Honourable Charles Stewart Rolls was born on 17th August, 1877. His parents, Lord and Lady Llangattock, were aristocrats and, like most aristocrats of that period, they had very little to do with their children's upbringing. Charlie had three older siblings, John, Henry and Shelley, and when he was born he joined them in the nursery on the top floor of the Llangattocks' London residence, South Lodges, where a very fierce Scottish nanny was in sole charge. Nanny Gilbert believed in a strict routine. The children were bathed every day at the same hour, fed four times a day and come rain or shine they were given regular outings for large helpings of fresh air around Hyde Park. Displays of affection towards any of the children were frowned upon.

Now and again Charlie and his brothers and sister were displayed to their parents in the drawing room at around five o'clock for approximately twenty minutes, a schedule of five minutes per child. Lady Llangattock took this duty of inspecting her children regularly very seriously. Lord Llangattock on the other hand enjoyed these rare occasions and always wished he could see more of his children. When Charlie was a year old, his siblings, at the ripe old ages of eight, seven and five, were packed off to boarding school while he was sent to the Llangattock country estate, the Hendre, near Monmouth in Wales, where it was felt he would get a much better quality of fresh air. Charlie had almost learnt to walk and was extremely adept at moving on all fours with the greatest possible speed, preferably away from Nanny Gilbert. As yet, he had not learnt to smile and viewed his world through his large, slightly protruding eyes with disconcerting intensity.

The Hendre was originally built as a shooting lodge at the

beginning of the eighteenth century and then extended in Victorian times into a huge red-brick mansion in the fashionable style of a Norman castle. Although the exterior looked magnificent, the interior was disappointing, with small rooms and narrow staircases, but the stables and outhouses were extensive. Lord Llangattock was a keen huntsman and kept an extensive livery stable. Although Charlie was not in the least bit afraid of horses, even as a toddler, he was never really interested in them. He was only interested in the wheelbarrows that the gardeners habitually used to transport tools, plants and rubbish from one spot to another. These long-suffering and very kindly gardeners spent a great deal of time carrying him around the estate to cries of "faster, faster!" Naturally they felt sorry for this lonely little boy, but they were also fond of him and intrigued with his inventive games and his fascination with all the garden machinery and their functions. If Charlie's small hands could disassemble a particular tool, he would be in his element, and the gardeners would indulge his curiosity and show him how the tools worked. As he grew out of the toddler age he moved on from wheelbarrows to bicycles and was in absolute heaven if he could persuade one of the men to hoist him up on to any bicycle belonging to them as a passenger, where he would perch precariously on their front crossbar, crying "faster, faster!" One day he found an old bath chair in one of the sheds, hauled it to the top of the long, winding drive, jumped in on the slope as it gathered speed, and careered down the hill, gauging the exact moment when he had to jump free before being thrown out and run over by the chair itself. Naturally there were plenty of accidents but the gardeners were astonished at his fearlessness; they had learnt that it was pointless to try to stop him from such dangerous activities. It was quite beyond the gentle gardeners to divert Charlie from anything that he was determined to experiment with in the form of transporting himself from one point to another without his feet touching the ground.

Unfortunately, Charlie was the despair of his tutor, who could find nothing to fire his imagination or attract his attention in any way whatsoever. Charlie simply switched off and daydreamed about returning to his friends the gardeners as soon as the tedious tutor gave up the unequal struggle. At the age of seven the Llangattocks decided to send Charlie off to a small traditional prep school, Mortimer Vicarage School, where the education was less rigid than the established preparatory schools. Charlie was just as unenthusiastic a pupil and equally unenthusiastic on the sports field. He lived for the holidays and his return to the Hendre where he could experiment with his tools and ride the gardeners' bicycles.

Strings were pulled and he was sent to Eton, but still he did not show any promise. He was a solitary boy with very few friends; not that he was ever bullied, he just had nothing in common with any of the other boys. They appeared to Charlie to have no interests other than sport, which he felt was pointless. Charlie had a sharp, caustic tongue and a slapstick sense of humour which forestalled any tormenter who thought they could get the better of him. Frankly he was bored with the subjects he was supposed to study and bored with the company he was forced to keep; as usual he longed to get back to the Hendre and do something useful.

"Rolls, I regret to say," complained Eton's Headmaster, "is still forgetful and irregular, sometimes even to vanishing point! And moreover at times, his thoughts are far away in dreamland or vacancy, instead of being with his work."

Charlie decided at the age of sixteen that the only way to escape the utter boredom of the classroom and tediousness on the sports field was to join the College Volunteers. This was a very surprising decision in view of his solitary disposition, but even that did nothing to detain him at Eton and after three years he left. Nothing that Lord Llangattock said could persuade Charlie to stay, and indeed he knew that if he forced him to stay Charlie would no doubt make sure he was expelled. Already, Lord Llangattock was aware of Charlie's strength of character and he had a strange faith in Charlie's ultimate destiny.

Of course, once Charlie was back at the Hendre he was anything but idle. He immersed himself straight away in his workshop, which he had created near the stables. Charlie's reading material had never coincided with the school curriculum. On his own, at the Hendre, he pored over newspapers and magazines, learning of new technology. It was in one such newspaper article that he read about electricity and immediately set about finding out how it could be installed in the Hendre. He investigated for himself how it worked and what the advantages were. He tried to persuade his father to install it but without much success until he pointed out the financial advantages. Lord Llangattock was notoriously parsimonious, and given his curious faith in his youngest son, despite the appalling reports on his academic work from Eton, he could see in him the conviction and confidence that he possessed when he worked on a new idea or invention. So, backed up by the possible financial advantages that Charlie had pointed out, Lord Llangattock agreed to allow Charlie to install a small electric plant. Charlie and his best friend the head gardener, and the rest of the gardeners, worked tirelessly until they finally succeeded in getting the machine to operate. In 1893, aged sixteen, Charlie had supervised the first complete electric installation in a private house in the county of Monmouthshire.

At this point in his young life Charlie thought that the electricity industry was going to be his chosen career, but since he had taken up bicycling as a fairly serious sport he thought that he might be able to combine the two.

At last he could see a possible career ahead. His tutor, who had been appointed by Lord Llangattock after Charlie had left Eton, had been struggling to get him to concentrate on academic studies, suddenly had his attention when he asked Charlie outright what his real interests were. Charlie's reply that he was interested in the creation of making electricity work in different forms and his curiosity in the mechanics of cycling led the young tutor to point out that if he wanted to pursue these interests he might be best suited to studying engineering at Cambridge. The goal of learning engineering at Cambridge galvanized Charlie to concentrate on subjects he had previously found boring. He made an extraordinarily determined effort to study the subjects that would allow him to sit the Cambridge entrance exams. Charlie knew he had a great deal of catching up to do. Encouraged by Charlie's efforts, the tutor told Lord Llangattock that with some serious extra coaching Charlie could well scrape a pass into Cambridge. Suddenly, Charlie realized that his future depended on this challenge. Unused to academic study but familiar with research into subjects that interested him, he applied himself with a zeal that astonished his tutor. He was rewarded with a commendable pass with which he managed to squeeze in as a late entrant. There, Charlie settled down to study the only subject that interested him, engineering, and he took part in the only sport that he was any good at, cycling.

His father was deeply impressed and proud that Charlie had finally achieved a place at Cambridge. As a reward for his hard work, his parents brought him along on a grand tour of Europe during one of his vacations. Lord Llangattock was a keen yachtsman and they visited many wonderful places, sailed through the Mediterranean to the Crimean battlefields, visited Moscow and toured Italy and France. Charlie spent most of his time down below, taking his share of the watches and studying the mechanics of the engine, but he realized nautical engineering was not for him.

During one of Charlie's vacations Lord Llangattock took him to Paris, where he saw his first car and a seed was sown. At the time he was still heavily involved with cycling events and although he was never particularly successful in the races he entered, he was very taken with the mechanics of the sport and wrote about the techniques of cycling for cycling magazines. By now he was in his element at Cambridge and had bought himself a typewriter so he could write to magazines on the subject of cycling. Interestingly he wrote to two brothers in America who had set up a cycling shop in Dayton, Ohio and were equally interested in the mechanics of cycling. Of course, they were the Wright brothers.

Charlie was fascinated with all aspects of engineering and attended any workshop where he felt he could benefit and learn new ways of introducing fresh inventions and ideas into the present day.

This was not all that Charlie was introduced to at Cambridge. Charlie had led a very monastic, male-orientated life until now. He had had virtually no relationship as such with his mother and he had seen very little of his sister Shelley, over the years, but suddenly he came into social contact with "girls". He was a very handsome young man and the girls were attracted to his tall, dark and handsome looks. His slightly protruding dark brown eyes seemed to examine and weigh up a person almost before they had been introduced. Some people found this rude; others found it compelling. Rather bewildered at first by the attention from the opposite sex, he was hugely flattered and did not really know how to deal with it. The May Balls were looming, and he had no idea how to dance, so with typical determination not to be made to look foolish, he embarked on dancing lessons. By the time the season arrived, he was quite proficient and had plenty of potential partners aching to become his. With all this attention and flattery he suddenly found he was able to turn on a creditable performance of immense charm, to great effect, and he very soon realized that this was an extremely useful asset. Charlie's mind was never idle. He was always preoccupied creating something new and imagining and exploring possibilities, and very often when he was carrying on a social conversation his mind was totally occupied elsewhere so that he appeared to be rather cool and aloof, and inevitably he sometimes gave the wrong response. Some people were discouraged by this aloofness and apparent rudeness; but women on the other hand seemed to find it attractive. Anyway, the long and the short of it was that he became an unwitting social success.

Weekend house parties were a common social activity, and one weekend the son of Sir David Salomons invited him to his family home, Broomhill, near Tunbridge Wells. Sir David took an instant interest in this aloof young nineteen-year-old friend of his son's from University. He instantly saw a young man, not only of vision, but also of drive and dedication to his future engineering career. In 1896 there was virtually no motoring in Britain and only a handful of men owned motor vehicles. Sir David was one of them. It was at this weekend house party that Charlie met Vera Hedges Butler and her father, a very rich and successful and also very jolly wine merchant, Frank Hedges Butler. Apart from the fact that Vera was very beautiful Charlie was also attracted and deeply impressed by the fact that at the age of sixteen she could handle a motor car. During the weekend Vera and Frank took Charlie for a ride around the nearby countryside and told Charlie that they would be delighted to bring him along on their next motor rally. Of course Charlie was thrilled. He thought Vera was simply wonderful and Frank was one of the nicest and jolliest people that he had ever met. Charlie wrote to his father in a high state of excitement, extolling the wonders of the motor vehicle and how his new best friends could get up to the amazing speed of 20 mph. He told Lord Llangattock that he was definitely going to get one of these autocars, and that he was saving hard to do so.

In 1899 the Boer War was declared and Charlie decided that he would volunteer and offer his services, firmly believing that he could operate a tricycle over South Africa's difficult terrain and help communications where petrol was so scarce. He could be a courier between army posts. When Lord Llangattock heard of Charlie's plan, he was horrified that his favourite son might go to South Africa with such a mad, dangerous scheme and offered to pay him the equivalent of the Queen's Commission allowance providing he dropped the whole idea.

Charlie was always made aware by his parents that "money did not grow on trees". They were extraordinarily careful in saving money where they could. There was no doubt that Charlie inherited this trait twofold, and gave parsimony a new meaning to frugality. It wasn't that he didn't pay his share but he always made sure that his share was a lot less than anyone else's. For instance, he would join his friends for a meal and while they might eat three courses he would have sandwiches. Nevertheless, he was immensely grateful for his father's generosity and his faith in him. Once when his father offered a further loan to tide him over, he refused to accept it unless he could pay his father a proper rate of interest.

Charlie weighed up the pros and cons of his father's offer and coolly decided that this allowance would enable him to start his own company, C.S. Rolls and Co. His mother was appalled that a son of hers was actually going to go into trade and vowed that she could never speak of it or have anything to do with such a ghastly decision. "Trade!" Charlie was quite unmoved by his mother's attitude, confident that his father would always give him moral support, although not necessarily financial. Charlie was the ripe old age of twenty-two when he set up his company. The event was recorded in the Daily Mail by a young journalist called Charlie Hands.

Charlie set up his business with a determination and entrepreneurship that was admired and scorned in equal measure. Most of upper-class society were disparaging: the conservative, old school population could not bear to contemplate change while the emerging adventurers followed every move of the new autocar owners' escapades. They turned out to cheer the motorists on and gradually they became well known, while the general public slowly began to abandon the horse and cart and buy motor cars of their own.

Charlie Rolls became a celebrity and was always in the news. In 1900 the small band of motoring pioneers were desperate to promote the motor car more widely. On the "Thousand Mile Rally" to promote the motor car Charlie, Vera and Frank Hedges Butler teamed up. Charlie loved their company; they were quite unpretentious and fun. Charlie was

The Hon. C.S. Rolls. (Photo by courtesy of the Royal Aero Club)

quite unused to fun but with Vera and Frank he could relax. Frank was so good humoured and easy going and despite the age difference they were to remain great friends for the rest of Charlie's life. He was a perfect companion, especially when Charlie became "strung up" and autocratic when things were not going quite his way. The fact that Vera had totally mastered the Renault owned by her father was a source of deep admiration on Charlie's part. Vera was a shy girl but her interest in motor cars was almost as intense as Charlie's. He loved having her by his side. Unlike other girls she didn't set out to challenge him or seek attention but was always interested in everything they did together. In 1900, at the age of nineteen, Vera and Charlie were inseparable but perish the thought that they declared undying love to each other. In 1900 an aristocrat never fell in love with a girl from a "trade" background – actually an aristocrat never fell in love, and a girl from a "trade" background could never have aspirations of reciprocal love with an aristocrat in any circumstances, however mutually compatible or attracted to each other they were.

Nevertheless, in 1900, Charlie, Vera and Frank attended the Paris Exhibition. Charlie valued both Vera's and Frank's knowledge of French motor cars and Charlie was keen to stock his C.S. Rolls showroom in Conduit Street with some impressive motor cars. Frank was used to driving on the Continent and introduced Charlie to a choice of manufacturers. France was way ahead of the British in all aspects of motoring.

As far back as 1893 France was the first country in the world to introduce a driving test, along with the first vehicle registration plates and parking restrictions. The first driving licences were not introduced in Britain until 1903. During their visit to the Paris Exhibition Vera took her driving licence and passed her test there. She was the first British

Miss Vera Hedges Butler – Founder of the Aero Club – subsequently The Royal Aero Club. (Photo by courtesy of the Royal Aero Club)

woman ever to do so. Of course, much to Vera's delight and embarrassment Charlie, over the moon with excitement for her, kissed and hugged her warmly, which surprised Charlie as much as Vera, and quite overwhelmed him, completely unused as he was to any display of emotion.

By way of celebration Vera suggested that the three of them make an ascent over Paris. Charlie had never been in a balloon before. It was a beautiful day and the sights of Paris were spread before them as they floated past the Eiffel Tower, over the Bois de Boulogne, the Arc de Triomphe, the Champs-Elysées and Notre Dame which quite took Charlie's breath away. Charlie was hooked and he asked Vera if she would do him the honour of becoming his ballooning partner. Vera smiled shyly and was thrilled, naturally.

In September 1901 Frank Hedges Butler, Vera and Charlie had made plans to go on a motoring tour, but at the last minute Vera's Renault 4 caught fire and the tour had to be cancelled. Instead, rather humiliated, Vera arranged a balloon ascent with Charlie, her father and the celebrated aeronaut Stanley Spencer as pilot, all members of the Automobile Club. The date was 24th September, 1901 and the ascent was made in the grounds of the Crystal Palace. In a balloon basket floating peacefully across the sky over Kent, twenty-year-old Vera proposed the formation of an Aero Club and, uniquely for that particular period in history, it was agreed that "membership should be open equally to ladies and gentlemen, subject to election". Exhilarated by the flight, and motivated by a bottle of champagne supplied by Frank, they all toasted the formation of the Aero Club proposed by Vera.

CHAPTER TWO

The Shorts

It so happened that three brothers, Horace, Eustace and Oswald Short, had also attended the Paris Exhibition, Horace to demonstrate his Auxetophone on top of the Eiffel Tower, which deafened the whole of Paris, and Eustace and Oswald to examine the construction of French balloons.

Horace was the eldest of the Short siblings; there was also Ernest, followed by Eustace, followed by Alice and the youngest, Oswald. Horace was born on 2nd July, 1872 and when he was a toddler he suffered a head injury which developed into meningitis. Unfortunately, as a result of the head injury and possibly because of the meningitis, Horace was left with an abnormally sized head and a rather frightening appearance. It was clear that his brain had not suffered and it soon became evident that he had an exceptional intellect. By the time he was fourteen Horace was bored with school and knew more than his teachers, so not only did he beg his father to take him on as an apprentice but so did his teachers. There was nothing more the school could teach Horace and he was too young to go on to higher education in those days. By now the family had moved to Derby where their father, Samuel Short, was the chief engineer to the Stanton-by-Dale ironworks. Naturally his father's foreman was unhappy about taking on such a young apprentice but Horace soon learnt the ropes and by the time he was seventeen he had become a skilled engineer. Once again Horace needed the challenge of something new. Frustrated with the restrictions of the limitations of the community in Stanton-by-Dale, Horace decided he should broaden his horizons and see more of the world. For an eighteen-year-old, he was extremely mature, physically strong, an accomplished engineer and quite fearless. Horace was also blessed with a wonderful sense of humour, which often got him into trouble, but was a gift in the years to come that was almost as valuable as his extraordinary intellect. In 1890, the eighteen-year-old Horace made his mind up to travel to the other side of the world and visit an uncle in Australia who had emigrated to Newcastle, New South Wales and had become a mining engineer.

Eustace, who was born in 1875, was three years younger than Horace. He was a quiet boy who was not impulsive; he considered his actions carefully; he appeared to be shy and yet had no fear of speaking up when necessary. He worked hard, was surprisingly single minded and was very accomplished on the sports field, yet he did not stand out from the crowd. Eustace's and Horace's personalities had always clashed, which even as boys led to heated arguments, but despite Horace's dominating character Eustace could stand his ground and did so, much to Horace's frustration.

Oswald, on the other hand, the youngest in the family, born in 1883, eight years younger than Eustace, and eleven years younger than Horace, turned out to be a much more balanced character than either of his brothers. As a little boy Oswald idolized Horace, and followed him around like a shadow. For a teenager, Horace was very patient; he would perform little magic tricks for Oswald and helped him build forts and castles and kites. He would take him for walks across the dales and fly the kites they made together and tell him stories about the animals they came across on their walks. Horace affectionately called Oswald "The Kid". When the time came for Horace's departure Oswald begged Horace not to go and asked him "Why do you have to go?" Horace tried to explain as best he could that he had to go on this big adventure to see what was on the other side of the world, and he would write to tell him all about it, but he would return. He told Oswald to work hard at school and learn all he could. At seven years old Oswald tried to be brave and tried to promise to do as Horace said.

There were two other siblings: Ernest, a year older than Eustace, and Ethel (who was called Alice by the family), two years older than Oswald, but they played no part in the future of Horace, Eustace and Oswald.

When Horace left the family, he made his way to Liverpool and was taken on as part of a ship's crew on a freighter bound for the Far East. Naturally, he had never been to sea before although he had often boarded the freighters at Newcastle when they were in dock and found a friendly sailor to show him around. As a boy he had taught himself to swim, and as a teenager he would often win races in the local baths. He was confident he could handle whatever disaster might occur at sea. Little did he know that he was going to encounter several. He acclimatized quickly to the conditions on the ship and by the time it docked in Hong Kong he was like an old hand. He stayed in Hong Kong for a few weeks and was tempted to travel into China, but a place on a freighter bound for the South Seas turned up, so he boarded that for the next leg down to Australia. Unfortunately, the ship was hit by a typhoon and sank, scattering the crew. Horace clung on to a piece of timber which washed up on a remote island. The vision of Horace emerging from the sea was an awesome sight for the islanders, who appeared to be half his size. They stood in a small group, their spears held menacingly, poised to bring this apparition down. As luck would have it a wild pig broke through the undergrowth on to the beach where the natives had gathered, and with great presence of mind, Horace unwrapped his pistol from the oilskin he had tied around his waist and shot it. Horace let out a roar of triumph which scattered the natives in terror. When they crept back to the beach they found Horace dramatically posing with one leg on the pig, pointing to the dead animal. The natives cheered, Horace cheered and they carried him and the pig shoulder high into their village proclaiming him as a god. It certainly saved his life but he was then their prisoner for months before he could escape. He wrote to his family of his final escape.

"I made several attempts to get away, but they all failed," he wrote. But eventually he eluded them. "I woke up one night when all were asleep, and rolled myself over inch by inch, towards the outside of the house, snoring like a pig. At last I got outside. I stood up for a second or two to see if all was quiet, and then I dodged quietly through outside, then I set off running, and never felt the ground I went over till daylight. By the time I had put twenty miles between me and the village, but still it was not over safe, so coming across a canoe on the beach, I slipped into it, and it being a small one, I could make about eight miles an hour in it. I set right out to sea, so that I could just see the land and paddled all day till sunset. You see, I had my compass, and I knew the position of all the islands by heart. I landed at night and left the canoe and next day I made off on a west by north west course, knowing that it would bring me to the little settlement where Europeans lived. About thirty Germans, two Swedes and six Englishmen comprised the whole of the white people on the islands. They all lived in this settlement and were afraid of venturing more than a mile out of town. They were terribly astonished when they heard where I had been, and said that it was a complete miracle that I had ever got back alive. They told me that the natives were nearly all cannibals, and that one or two white men who had ventured out to see what was on the island had been murdered and eaten. At last I got on board a ship and sailed for Sydney, which was about 3,000 miles away, and here I am safe and sound."

The letter was published in a Chesterfield newspaper in 1893. Horace had then been out of touch with his family for a long time and did not he know that his father had died of consumption only a year or so after his departure in 1891, leaving the family penniless. The family had no way of contacting him. Horace had left Australia by now and found a passage on another freighter. He called in to Cape Town but did not stay long and finally reached Mexico, where he talked his way into running a silver mine. He proved to be very successful and ran the mine very profitably, albeit that by then he was only twenty-one years old. Fortunately Horace had an incredibly keen ear for languages. Already a fluent French speaker, he learnt Spanish in no time at all.

It was some time after he got to Mexico that the family knew roughly where he was. Naturally he did not think to tell them that he lived in bandit country, where life was cheap, and that he was constantly on his guard against thieves and murderers. Why should he have done, since the last thing he expected was a visit from his brother Eustace?

Back in England, the family were in dire straits and even Oswald, then only ten years old, took a job as an errand boy to try to earn some money to help support his mother. Eustace was beside himself, unable, it seemed, to make contact with Horace to tell him of the plight they were in. So scraping enough money for the journey to Mexico he decided there was nothing for it but to find Horace and tell him to come home, do his duty and support the family.

For someone who had barely been out of Cheshire it was not only very naive of Eustace to embark on such an expedition but actually very brave. Landing in Mexico, he had little idea of where to go or how to get there. Not a good start, especially when he hired a guide who relieved him of most of his precious money and his return ticket. He got no sympathy from the local police, because robbery was all too common, and in any case they could not understand a word he said. Fortunately, in the end, they called up the local British Consul who came to rescue Eustace and helped him discover exactly where the mine was where Horace was meant to be working. The only address Eustace had was the name of the mine.

Three days later, by mule, with a so-called trusted guide and within only five miles of their destination, the guide abandoned Eustace to fend for himself in sweltering heat in dusty, arid and dangerous rocky terrain amongst what appeared to be wandering bandits threatening to ambush them at any moment. Within a very short time Eustace was indeed accosted, and exhausted, hungry and extremely irritable, ignoring their threats of extortion and violence he shouted furiously at them "to shut up" and demanded them to lead him to Mr Short. To his astonishment, their whole attitude changed as soon as they understood that he was Senor Short's brother. They offered him food and water and escorted him like royalty to Horace's mine. There, they knocked timidly on the door to Horace's office and ushered Eustace in with a great deal of bowing and scraping.

Horace was astonished; he could hardly believe his eyes but did not seem overjoyed to see his younger brother.

"What on earth are you doing here?" he shouted. "Don't you realize this part of the world is extremely dangerous, there are bandits out there who could have robbed you and then left you for dead . . ."

"Don't you lecture me," retaliated Eustace furiously "I've crossed the Atlantic, I've already been robbed of everything and beaten up, and with nothing more than a water bottle I've come across what must be the most inhospitable and barren land in the whole of the world, and I haven't done this just to be lectured by you." Before Horace could intervene Eustace took a deep breath and let fly another tirade. "Had you bothered to keep in touch with your family it would have saved me a great deal of trouble. How else was I supposed to contact you? Our father is dead, he died of consumption some months ago and the family are now in desperate straits and as you are now head of the family it's your duty to return to Cheshire immediately and look after our mother!" Eustace promptly collapsed on the nearest chair.

It was difficult to tell whether Horace admired Eustace's determination and extraordinary intrepidness, or thought that he was incredibly stupid to make such a dangerous journey. Nonetheless he was not ready to return to Cheshire. There was no question of him leaving his silver mine at that point in time. Instead he gave Eustace £500 to tide the family over and suggested that he invested most of the money in a coal merchant business. He then dismissed Eustace, giving him his fare home and an escort back to the port. Eustace was furious at being dismissed so swiftly but at the same time relieved and grateful for what was then a huge sum of money.

One very good reason that Horace was unwilling to leave was that he had just invented a device which he called an Auxetophone which amplified sound to an extraordinary level. He had invented this device to use to contact miners below ground and was naturally preoccupied with developing it. He realized that the potential for this device was enormous and had been greatly encouraged by a Colonel George Gourmand, a former agent for Thomas Edison, who believed that Horace should consider marketing it.

Oswald was bitterly disappointed that Horace had not returned with Eustace. Eustace assured Oswald that everything would be fine now and Horace would come home soon.

It was shortly after Eustace's return that the two boys witnessed a free-flying balloon advertising a fair at one of the local towns. The balloon floated over them for long enough for them to catch where it was being held, and off they duly went the following week. When they got to the fair the same balloon hovered over them. Both Oswald and Eustace were enthralled. It certainly beat flying kites. There was a stall selling toy balloons which they were immediately drawn to and they bought several. Eustace thought how wonderful it would be to fly in a balloon while Oswald wanted to make one.

The boys spent hours carefully dismantling the toy balloons they had bought and recreated the panels by cutting thin newspaper, experimenting all the time to achieve a perfect model. Eustace devoured articles on balloonists making ascents. Oswald sought out technical and scientific journals and they both scoured the newspapers and magazines for any information connected with ballooning. Oswald even studied his father's engineering pocket books for any useful information that could help. Despite the limitations of his education Oswald's meticulous attention to detail and hunger to learn, even at such a young age, allowed him and Eustace to achieve some success.

Coal merchants in Cheshire were numerous. Eustace decided they could make a better living in London. By the time Horace returned in 1895, the family had moved there and Eustace had already set up the coal merchant's business to take care of the family. Horace did not stay with the family for very long. Once he was satisfied that all was well he moved to Hove and with Colonel Gourmand's support set up a studio to work on his Auxetophone, which, on the latter's advice, he patented. At the same time he offered Eustace and Oswald space in Hove to develop their ridiculous ballooning interests.

In fact Eustace's interest in ballooning had deepened and he desperately wanted to make an ascent, while Oswald was still more interested in the dynamics of ballooning. With the toy balloons they had bought originally Oswald spent hours carefully reconstructing models and experimenting with the way their dynamics worked. In London they had better access to information on the current meetings and gatherings where balloons were all part of the entertainment. Eustace soon got to know the regular balloon enthusiasts and eventually his curiosity and his eagerness to learn overcame his shyness to strike up a conversation with them. In 1897 he plucked up courage to introduce himself to a Mr Griffith Brewer, a well-known and respected balloonist who attended these meetings regularly. Griffith Brewer eventually took pity on Eustace and invited him to make his first ascent with him. Eustace described the ascent as a "rough voyage", with the balloon splitting open on a tree on landing, but this in no way diminished his enthusiasm.

The brothers managed to scrape together £30 and bought a Spencer-built balloon, called *Queen of the West*, which they trailed around at fetes and fairs giving rides to anyone who would pay a fare. It rattled as they filled it and was blown and worn, and very often the fabric cracked as they filled it so they had to stick it together with sticking plaster. However, Griffith Brewer sportingly accepted a ride in it and after a great deal of persuasion Horace also agreed to make an ascent in it. After this momentous occasion a newspaper article appeared describing this "very special and expensive novelty, [which] was both figuratively and literally a big undertaking", but the results financially at the annual Teddington Church Temperance Guild also showed that it was a "big draw", even though the weather apparently was appalling, gloomy and depressing. The report was as follows:

"Under the superintendence of Mr E Price, manager of the gas-works, and his foreman, Mr C. Douglas, the main in Vicarage Road was continued into the first meadow, and on to the end about a hundred yards of canvas hose was connected which reached to one of the further meadows, where the balloon received the 35,000 cubic feet of gas. The process of inflation commenced about 12.30 and from time to time, till five, the hour appointed for the ascent, an ever increasing crowd circled round the scene three and four deep, and watched the proceedings with great interest, two photographers occupying exalted positions with their cameras, with the object of obtaining a permanent record of the novel scene. Mr Short and two companions, who were to accompany him on his journey, superintended the operations, and gave minute attention to every detail. The crowd was a very orderly one – they always are at these fetes – and made the duties of the two policemen almost a perfect sinecure. Everything being in readiness for the start, the bags of ballast which held the monster to Mother Earth were removed one by one, Mr Short jumped into the car, and his two fellow voyagers into the ropes, and then at a given signal, the last tie was loosed, and amid a shout of applause, the golden tinted balloon shot into the air like an arrow from a bow, leaving in its train thousands of leaflets which came fluttering down from the car. In almost less time than it takes to write, the balloon was lost to sight, having entered the thick low lying curtain of cloud which overhung the scene. And then the crowd dispersed about the meadows to enjoy themselves in other ways . . ."

Eustace also wrote a report which recorded, "Those in the car besides myself, Mr H.L. Short and A. Williams aeronaut of Battersea. The balloon let go at 4.45 and immediately ascended, and in about 35 seconds was enveloped in a very thick cloud, during our passage through which we could scarcely see our hands, and it took us about five minutes to get through it. When we did we found ourselves in a cavern formed by two clouds, one above and the other underneath. At the end of the cavern the sun was shining brightly. On getting out of this cavern the heat of the sun expanded the gas in the balloon, which caused the latter to rise and in a short time an altitude of some three to four miles was reached. At this point we became deaf. We then thought it advisable to come down, and let out a little of the gas for that purpose, which made the balloon drop to within a mile and a half of the earth, which could be seen through the opening in the clouds. The balloon drifted on, now up and down, til Otford in Kent was reached (not Dartford as had been stated, nor Oxford as had been reported). The balloon at this point dropped to within 100 feet of the earth, and we were then able to talk to the inhabitants of various villages, who told us they were about ten miles from the sea coast. After going along a little the balloon again rose, this time to an altitude of two miles at which height we skimmed along for a good few miles. After a time we heard sounds of hammering and the barking of dogs; fearing we might drift out to sea, we let out more gas, and came down nearly to the ground just on the edge of a wood, over which the balloon travelled for about half a mile through the tops of the trees. We threw out the grapnel when we got to the edge of the wood. This held the balloon for about half a minute. It then broke away, and every now and then kept taking a hold which caused us some nasty jolts and in fact once nearly pitched us out of the car. After we had got through the wood we came down on a large moor which was covered with big flint stones. By this time the balloon had lost so much gas that she was unable to rise, being perhaps only half full and the wind catching hold, carried it along at a very fast rate over the stones. At length the balloon and car lay on their sides on the ground. We waited quite an hour before we ventured to get out, and then we only got out with assistance, the balloon being just on the edge of another wood. The total distance in all covered was 43 miles (Oswald called it 45). When the balloon came down for the last time it only wanted a few minutes to six. After we got out of the car the balloon was packed up and taken in a cart to the station. We were entertained to supper by Mr Wellband, of Castle Farm, Otford, who said that when he saw us, the balloon itself was hidden in the clouds, and only the car could be seen speeding along."

The most significant upshot of that was that Horace had a complete sense-of-humour failure. He was practically speechless when the car finally came to rest on its side. When he recovered he swore that "they are nasty, utterly useless contraptions and I have no intention of ever, ever, ever making another flight and as for Eustace and 'The Kid' they are completely and utterly mad, balloon mad!"

Eustace and Oswald were not to be put off by their big brother and when Horace announced that he was going to attend the Paris Exhibition, Eustace and Oswald announced that they were going to accompany him. They realized all too well that their *Queen of the West* was old, rather dangerous and that the design was out of date. They had already taken the top floor of Horace's studio in Hove to construct their own balloon, and as usual they had done their own research meticulously, and had specifically come to see the extraordinarily well-designed balloons of the Astra Societé de Constructions Aeronautic of Billancourt. The days of hiring themselves out as showmen in the *Queen of the West* were drawing to a close, and they now had their sights set on manufacturing their own balloons. It may have been the boys' youth and enthusiasm that inspired the designers to give Eustace and Oswald the technical guidance they craved, but in any event the boys returned to England with their ambitions now firmly focused on manufacturing their own balloons.

When Eustace and Oswald arrived back in Hove two shocks awaited them. The first was that the brothers were absolutely dumbfounded when Horace introduced them to his new wife, Marie Claire, who was obviously pregnant. Marie Clare was a Frenchwoman from Jersey who spoke virtually no English. Horace had always been a "ladies' man" and it astonished the boys that their rather ugly brother had always seemed to have a string of ladies to choose from. Horace was a skillful flirt and knew exactly how to handle a woman, but a married Horace was something that neither of the brothers had bargained for.

Horace's second bombshell was that he had decided he had gone as far as he could with his Auxetophone. He had been searching around for a new project when he was offered a contract with Sir Charles Parsons to help develop his turbine engine. So, with great enthusiasm, he accepted, and therefore he and Marie Claire were going to move to Northumberland.

"Northumberland!" shouted Eustace. "I see, abandoning the family once again, for turbine engines!"

"Better than balloons!" shouted Horace in retaliation. "Where will they get you?"

Horace had got his new challenge, and so far as he was concerned, Eustace's and Oswald's venture was of no interest to him whatsoever, and frankly of little importance either. Eustace was incandescent with fury at being dismissed in such a way. Shouts of abuse between Eustace and Horace echoed over Hove for days, while poor little pregnant Marie Claire looked on in dismay and bewilderment. "The Kid", Oswald, took it on the chin, feeling that Horace simply did not understand the significance of what they were trying to achieve, and the possible road that he and Eustace had embarked on. He was sure that in due course they would need Horace, and Horace would in due course know just where the future was going to lead.

Horace duly disappeared to the north of England while Eustace and Oswald moved back to London, using their mother's home as a business address and renting the ground floor of a carriage shed in London Mews, London Street, just off the Tottenham Court Road. The mews and street have since been renamed Maple Place and Maple Street after the famous department store that stood nearby. The first floor of this premises was reputed to have been used by Whistler, the artist, but unfortunately when they came to rent it the premises was used by a musician who practised endless scales all day long.

However, armed with their new-found knowledge of better balloon making by their French counterparts, Eustace and Oswald began working in earnest to produce balloons that were superior in every way to their competitors in England. With all the enthusiasm and bravado of youth they set themselves up as:

SHORT BROTHERS
BALLOON MANUFACTURERS,
FLYING MACHINES,
KITES FROM PLAN,
AERONAUTICAL EXPERIMENTALISTS.

One day Oswald spotted an advertisement in *The Times* newspaper in which the War Office invited tenders for a contract to construct balloons for the "Government of India". These balloons were intended for reconnaissance work with the Indian Army. They worked round the clock to present their tender, giving detailed specifications as to the construction and manufacture of suitable balloons. Their hard work paid off, and much to Eustace's and Oswald's excitement and delight, they won the contract. It was their very first commission, and their very first aircraft to be sold. The hallmark of these balloons was high quality, giving the brothers an enviable reputation.

CHAPTER THREE

The Wright Brothers

On the other side of the world in Kitty Hawk, North Carolina, near the ocean, the Wright brothers of Dayton Ohio, on 17th December, 1903, flew their first powered heavier-than-air machine.

Wilbur and Orville Wright were the sons of a bishop. There were seven siblings altogether, although twins had died in infancy. Their very first interest in flying machines was when their father came home one day with a toy helicopter based on an invention by Leonardo da Vinci and the aeronautical pioneer Alphonse Pénaud. It was made out of paper, bamboo and cork with a rubber band to twirl the propeller. The boys played with it for hours and hours and endeavoured to make copies. In another continent, Eustace and Oswald Short's imagination of and fascination with flight came from a toy balloon round about the same time, and they also endeavoured to make their own copies.

The Wright family moved several times before they came to rest in Dayton, Ohio. Wilbur was four years older than Orville. He had a strong intellect, was athletic, popular and a natural leader. Just before he was due to go to Yale he suffered a very nasty accident on the hockey field when he was hit in the face with a hockey stick, lost his teeth and suffered a bad head injury. The injury caused a crippling depression, and he was unable to cope with any stress whatsoever. He withdrew from his formal education and became virtually house-bound, caring for his terminally ill mother and reading extensively from his father's library.

Meanwhile Orville, who had never really settled at school and had once been threatened with expulsion, and who was thought to be wild and out of control, finally dropped out of school. It was probably true that conventional education did not inspire Orville in the least bit, and a comparison could be made with Charlie Rolls. Right from the outset he wanted to make things, and even before he gave up school, with Wilbur's help he had made a printing press and set up a small business. Fortunately, at that point, Wilbur was able to throw off the lingering depression caused by the accident, and joined Orville in the printing workshop as editor of a weekly paper which Orville was publishing.

In 1892, and more or less at the same time, capitalizing on the new bicycle design of wheels both of the same size, which had developed bicycling into a new dimension, the brothers opened a bike repair and sales shop called the Wright Exchange Shop. It wasn't long before they began to manufacture their own bicycles and changed the name of their shop to the Wright Cycle Company. This enterprise helped fund their growing interest in flying machines. They had begun to experiment with gliders and read every article on aeronautics they could lay their hands on. By now Wilbur was very much the leader in their enterprises and their experimental inventions. At the end of the 1890s Wilbur wrote to the Smithsonian Institute requesting information and publications about aeronautics drawing on the work of George Cayley, Octave Chanute, Otto Lilienthal, Leonardo da Vinci and Samuel Pierpont Langley.

They began their experiments in mechanical aeronautical experimentation that year. They realized that one of the most important aspects of their experimentation was mastering the control of their piloting skills. Their years of experience within their bicycle business had given them a practical insight into the necessity of balance and control. Their piloting skills were largely developed through practice with their bicycles and their gliders, learning how to master control and balance all at the same time. The fatal accidents of men such as Lilienthal and Percy Pilcher were lessons they also used to adapt their own methods to prevent the mistakes these brave aeronauts had made. During this period they employed a shop worker called Charlie Taylor, who became a most important part of their team, and who helped them build their first heavier-than-air machine.

The Wright Flyer 1 was built with spruce, a strong lightweight wood, which they used to build their own wooden propellers, designed by themselves. They had a purpose-built gasoline engine built in their own workshops by Charlie Taylor, their shop mechanic, in six weeks since none of the engine manufacturers they approached could meet their specifications. The Wright/Taylor engine was a primitive version of a modern fuel injection system having no carburettor or fuel pump. Petrol was gravity fed into the crankcase through a rubber tube from the fuel tank mounted on a wing strut. It had a wing span of 40.3 feet and weighed 605 lbs with a 12 horsepower 180 lb Engine. The Flyer cost less than a thousand dollars in contrast to the more than 50,000 dollars in government funding given to Samuel Langley.

In camp at Kill Devil Hills, North Carolina, on 17th December, 1903, after weeks of delay caused by broken parts in preliminary flying attempts, the first flight was achieved by Orville of 120 feet in 12 seconds. The next two were by Wilbur and Orville respectively of 175 and 200 ft. Orville's account of the final flight of the day related that:

> Wilbur started the fourth and last flight at just about 12 o'clock. The first few hundred feet were up and down, as before, but by the time three hundred feet had been covered, the machine was under much better control. The course for the next four or five hundred feet had but little undulation. However, when out about eight hundred feet the machine began pitching again, and, in one of its darts downward, struck the ground. The distance over the ground was measured to be 852 feet; the time of the flight was 59 seconds. The frame supporting the front rudder was badly broken, but the main part of the machine was not injured at all. We estimated that the machine could be put in condition for flight again in about a day or two.

Five people witnessed the flight from the US Government coastal life-saving crew: Adam Etheridge, John T. Daniels (who took the first flight photo using Orville's pre-positioned camera), Will Dough, businessman W.C. Briskly and a teenage boy living in the area called Johnny Moore. Sadly, when the men were hauling the Flyer back from its fourth flight, a gust of wind flipped it over several times despite the crew's attempts to hold it down. It was severely damaged and it never flew again. It was subsequently restored years later by Orville, and finally came to rest in the Smithsonian Museum.

At the end of that auspicious day, with great excitement the brothers sent the following telegram to their father:

> SUCCESS FOUR FLIGHTS THURSDAY MORNING ALL AGAINST 21 MPH WIND STARTED FROM LEVEL WITH ENGINE POWER ALONE AVERAGE SPEED THROUGH AIR THIRTY ONE MILES LONGEST 57 SECONDS INFORM PRESS STOP HOME CHRISTMAS
> (17th December, 1903)

Unbelievably the *Dayton Journal* refused to publish the story, saying the flights were too short to be important and in any case who on earth would be interested in such an event. Various inaccurate stories were leaked to the press subsequently but they were so ridiculous and so dismissive and inaccurate that Wilbur decided to issue their own press release. Of course, no one had ever heard of them or was the slightest bit interested in either them or their achievement. Wilbur then released their statement to the international press and, despondently, virtually forgot all about it.

Well, the news did leak back to Europe, not only to the Aero Club in France, who treated the news seriously and heightened their ambition to match the Wright brothers' achievement, but also to the very few, but very select, members of the Aero Club in England, who also treated the news seriously and with a fair degree of excitement.

Tucked away in the Tottenham Court Road being driven completely barmy with the piano player's scales, Eustace and Oswald found the article in an obscure magazine. Once again their ambitions were fired, and they knew they were going to have to move, as the restrictive space was hampering their development, apart from the screaming irritation of the pianist's practising all day long and ruining their concentration.

CHAPTER FOUR

The Aero Club

As the instigator of the Aero Club, Vera undertook, initially, the duty of recruiting members, enthusiastically assisted by her father, Frank Hedges Butler. As seasoned balloonists, naturally they had a circle of friends they could call on and Frank, as a successful wine merchant, had one or two potential clients he felt he could approach.

One such client was Frank McClean, an elected Fellow of the Royal Astronomical Society and a Member of the Royal Institution. Although he had trained at the Royal Indian Engineering College, Cooper's Hill and subsequently spent three and a half years in the Public Works Department in India, he had returned to take part in the family business, and had already developed his interest in astronomy. An introduction by Frank Hedges Butler made between Charles Rolls and Frank McClean sparked McClean's interest in ballooning. He had a most attractive personality: he was warm, amusing, adventurous, generous and completely without guile. Charlie took to him immediately, which was unusual; Charlie never took to anyone straight away. It did not take Frank long to make his mind up to join the Aero Club.

C.S. Rolls and Co. in the early 1900s had plenty of challenges. There were very few garages and workshops who actually understood motor cars, and petrol itself was scarce. Charlie had set up a smart showroom in Conduit Street but repairs were carried out in Fulham where his secretary, Florence Caswell, gallantly tried to keep some sort of order with the accounts. Charlie's profile as a motoring pioneer and a successful rally driver was high, and Charlie Hands of the *Daily Mail*, who had always admired Charlie, reported very favourably; which was more than could be said of *The Times*'s sceptical reports. Charlie was seen as a dashing, elegant aristocrat, the epitome of a brave and sporting Edwardian gentleman. He was newsworthy and aware that the publicity was good for business. Society ladies were often seen calling at Conduit Street in the hope of seeing him and inviting him to some social function or other. If caught, Charlie would be charming and flattering but avoided any commitment. He was a superb actor because he despised the empty Edwardian society and viewed it with utter contempt.

One day a young man called at the Conduit Street showroom and Charlie immediately mistook him for a "society waster" and sent him packing before the young man could even open his mouth. The following day, the young man was back and, before Charlie could dismiss him again, the young man began:

"Now look here, I'm at Cambridge University studying engineering, and I would really like to work for you in my vacation. I'm a pretty useful mechanic and actually I do think what you are doing for the motor car business is absolutely marvellous and in fact I must tell you that I have admired you, the whole of my life."

In fact the young man was only sixteen at the time! Charlie was naturally flattered and very taken with his good humour. He knew he could always do with some help, but, as usual, he didn't really want to lay out any wages. So he said:

"Fine, er . . . very well, but of course I can't pay you anything."

"Oh," said the young man, "I don't want any money."

"Splendid," said Charlie, sighing with relief. "What's your name?"

"John Moore-Brabazon," replied the young man. "Call me Brab."

Charlie could barely conceal his disbelief that the young man appeared to be so thrilled at being employed by him, the notorious Charles Rolls, and was perfectly happy not to be paid any wages. Astonishing! Charlie told the young man to report to a Mr T.O. Smith at his Fulham Road workshop and offices.

Tommy Smith was a cockney from the East End of London. As far as Charlie was concerned Tommy was a "gem", as not only was he a gifted mechanic but he had a wonderful cockney sense of humour which matched Charlie's own caustic wit. They got on like a house on fire and were as close and loyal to each other as two peas in a pod.

John Moore-Brabazon was a born optimist, loved life to the full, and had the capacity to like, indeed love, almost every human being he came across. He had an ability to find the good side of and be sympathetic to the failings of a character without judging them. He was exceptionally good humoured, and very popular with people from all walks of life. He could be quite outrageous, but in a harmless sort of way, so that the only person in any real danger would be himself. He was extremely competitive in any sport, and there were indeed many sports which he participated in over the years. He was clearly very brave, although he always denied it, but never denied his fear in times of extreme stress.

Brab, as he was affectionately called by his close friends, from the start almost hero-worshipped Charlie Rolls. He recognized that Charlie was "the strangest of men but one of the most lovable". He could see that Charlie possessed a true prophetic vision, with a very sound knowledge of mechanics, and an empathy with machinery that was uncanny. Brab found him to be a loner, who refused to take part in any conversation unless it was particularly relevant to his own interests. Charlie's only fault, so far as Brab was concerned, was his extreme meanness, which was almost a source of amusement. He hated to part with hard cash, to the point that one day, travelling back from Paris on a train, he had bought a ticket in the second-class compartments while his friends who had bought tickets in the first-class compartments urged him to join them. When he refused, they offered to pay the difference, and he happily accepted their offer, despite the fact that he was better off than any of them. On other occasions, while on racing tours, he would often insist that he and his mechanics slept under the car on the pretext that there was no accommodation available, when in fact he was simply trying to save the cost of hotel bills. Brab never minded sleeping under the car but the dropping of oil and dirt from a racing car did not improve their early morning appearances. Not that that bothered Charlie one iota. He did not suffer fools gladly and with his rather crude sense of slapstick humour it was hardly surprising he did not have a huge circle of friends. But Brab did become one, Charlie's better qualities outweighing his rather challenging characteristics.

Charlie decided that he liked this ebullient young man, and introduced him to Frank Hedges Butler, who immediately invited him to make a balloon ascent with himself, his daughter and a General Colville. Frank's enthusiasm was catching and although he was not a particularly good aeronaut, his total self-confidence and complete lack of fear was an inspiration to a mere beginner. Brab's description of his first flight, or rather descent, was recalled by him:

> The drill when you came down was to throw out a grapnel, which was meant to anchor itself in a field or hedge and pull you up while you "valved", i.e. let the gas out of the balloon at the top. However, in this particular case the grapnel didn't catch, or perhaps it got caught in a bush which wouldn't allow it to hold; and I assure you that we went over a mile – crash, bang, crash, bang – in that manner. As Butler weighed a good seventeen stones and for the most part he was on top of me, that particular part of the trip wasn't the greatest fun.

Brab's second ascent was just as adventurous and he recalled:

> I had arranged through the Aero Club to go up at the Crystal Place with Spencer, who was the great aeronaut of the time, and one other member. When I turned up, the balloon was all ready and there was the great Spencer complete with beard and a yachting cap with a balloon on the front, and he and I and another man got in. Well, we'd got the weights wrong, because the balloon wouldn't go up with the three of us; it was quite obvious it would take only two. The point was, who was to give way. The other passenger turned out to be a very remarkable man and later a great friend of mine, Professor Huntingdon. He was certainly the most formidable figure I have ever met. He had a very gruff voice, a very large nose, and was by profession a metallurgist. I was always terrified of him; but at heart he was the sweetest man and I came to love him very much. He came up to me and in a very gruff voice said, "Have you ever been up in a balloon before?" I said, "Yes, once. Have you?" He said he had been up three times and added, "Do you think you and I could manage

> it together?" I said that I didn't know but we could always try. So off we went, much to the astonishment of Mr Spencer.
>
> We sailed off – missing the towers, thank God – and drifted towards the east of London. Now, the first thing you should do when you get up in a balloon is to let the trailing rope out. This is a rope about a hundred yards long and fairly heavy, and the function of it is to take the weight off the balloon when you come down, for of course the more rope there is on the ground it delays your forward velocity. You should get this rope out as soon as you can. We didn't, with the result that I had to call Huntingdon's attention to the fact that we were coming down rather quickly and that he ought to throw out ballast. But he wasn't quick enough, and we hit the top of a house in the East End and knocked off the chimney pots. By that time the ballast we had thrown out was enough to make us rise again; then we let the trailing rope out and eventually came down just south of the Thames. For a long time we sat in the middle, not knowing which side the wind would decide to blow us; but it was all very enjoyable. And that was my first contact with Professor Huntingdon.

Brab was soon recruited into the Aero Club and, later on, Charlie asked him to go on the committee. He was astonished but thrilled to be asked, and accepted the invitation without thinking twice. Gradually the Aero Club increased its numbers, but only by very few. Most people regarded the Club as a crackpot organization, and would have nothing to do with it, but Charlie Rolls' enthusiasm and extraordinary prophetic foresight attracted the most adventurous, the bravest and the most entrepreneurial in a society that was largely stuck in the nineteenth century.

Round about this time Charlie and Brab began to experiment making gliders, following the attempts being made to launch a successful model in Europe. The difficulty was that having made their models out of very thin balsa wood and tissue paper, they did not know where to launch them. The air, of course, had to be completely still. Charlie came up with the novel idea of launching them from one of the upper circle boxes in the Albert Hall. Using his considerable charm, which he was able to turn on whenever the occasion arose, he persuaded an usherette to let them in and allow them to launch their models down into the stalls during a time when no one was around. It was an interesting learning curve for both of them and gave them much food for thought.

Another loyal and trusted friend of Charlie and the Hedges Butlers was an experienced balloonist and engineer called Griffith Brewer, who joined the Aero Club in the early days of its formation. Charlie came to rely on Griffith when difficult decisions had to be made and also as a compatible and reliable companion, particularly when he visited Wilbur and Orville Wright in Dayton, Ohio and later in France. He was a quiet unassuming man, and intensely loyal.

While Eustace and Oswald were completing their contracts for the Indian Government, and as a result of their reputation for the very high standard of their work, they were visited by a Colonel James Templer, who was at the time in charge of the Services Balloon Factory and School of the Royal Engineers, and was also one of the country's most notable balloonists. Colonel Templer wanted to see the Shorts' work for himself and was immensely impressed with the quality and impressive workmanship. Oswald later recalled:

"He addressed us as boys. I suppose we looked like it; we weren't quite, I was in my teens. He asked us: 'Do you know any members of the Aero Club?' and we said, "No, but we have heard there is an Aero Club' and he said, 'I will tell them about you.'"

In fact, Colonel James Lethbridge Brook Templer, one of the founding fathers of British Aeronautics, changed the course of the Short brothers' lives. His interest in and approval of the work they were achieving was a major accolade, and an indication that they had a great future ahead. Of course, at this particular point in time they were struggling to keep afloat. The Indian Government were slow to pay the bills and cash flow was always a problem. The brothers continued to devour every paper and magazine they could lay their hands on to learn everything they could about the current situation regarding balloonists and aeronautics. When they stumbled on a small article in the *Daily Mail* about the Wright brothers' success, they endeavoured to find out all they could from whatever source they thought might have information. The general lack of interest in the public domain was a source of extreme frustration. But Oswald's diligence paid off, and from the scant details he acquired, he began to experiment and invent his own designs for a heavier-than-air machine.

Templer was true to his word. At the time, Charlie Rolls was longing to purchase his own balloon and had asked his long-suffering secretary, Florence Caswell, to find him a reliable balloon maker. Charlie had already investigated several manufacturers and interviewed them, some of which turned out to be toy balloon makers. Templer pointed him in the direction of the Short brothers.

They were still in the cramped premises of the ground floor of the carriage shed off the Tottenham Court Road, but Charlie Rolls, nevertheless, sought them out. Although Charlie was surprised at the brothers' obvious youth, and although he thought Oswald was only a teenager, this in no way put him off, remembering his own early excursions at the age of sixteen when he installed electricity into the Hendre, nor did the rather humble premises deter him. Charlie Rolls was never one to be impressed by smart appearances, but he was deeply impressed by the extraordinary quality of the superb workmanship and the enthusiasm of the brothers. They were undoubtedly on the same wavelength as himself. He ordered two balloons on the spot. Quite out of character from his usual thrifty nature. He called one of the balloons *Britannia*, which he entered into the big Gordon Bennett balloon race. Miss Florence Caswell nearly fainted when she heard he had bought not one, but two balloons. She recalled Charlie Rolls' elation when he returned to the office with the news that he had found these superb balloon makers called Shorts, who would welcome any amount of orders. Oswald was terribly grateful to Charlie, not just for his custom but for the valuable advice Charlie gave to them. In Oswald's own words:

"By his advice and by his actions he saved us from financial losses, which, in those early days, we might not have been able to recover from."

The fortunes of the brothers began to change as other members of the Aero Club, such as Griffith Brewer, Professor Huntingdon, Cecil Grace and Warwick Wright also ordered balloons. But as the orders flowed in the piano scales on the first floor suddenly became more than either of them could stand. Fortunately they had found alternative suitable premises under the arches at Battersea Park. At first Eustace grumbled that no one would ever find them, but Oswald ignored his concerns, convinced it was the right thing to do. After all, it was convenient for Eustace's coal merchant's business, and they were close to the gas works to buy gas for their balloons. They also had far more space, and easier access to load and unload their spare parts and completed balloons, and, best of all, there were no piano scales to spoil their concentration. The only thing that Oswald found difficult to come to terms with was the proximity of the Battersea Dogs Home. Oswald adored animals and the thought of these abandoned animals always made him terribly sad; he had already adopted two of them. Eustace was determined that Oswald should not adopt any more.

CHAPTER FIVE

Charlie Rolls meets Henry Royce

Very early one morning, Rolls' Lillie Hall "Showroom", if you could call it that, in Fulham had become so overcrowded with cars that it was impossible to move. Charlie was losing patience trying to manoeuvre and shunt them around to make more space. Always an early riser, he had been up since dawn trying to improve the look of the overcrowded showroom. He climbed out of the last Packard in the line, swinging his long legs over the door, and carefully jumped down in the narrow space between the two cars, since there was no room to open the driver's door. He edged his way through the motor cars, pulled back the sliding plate glass shop front window to allow enough space to get out and walked across the street to view his work.

He noticed out of the corner of his eye a little knot of bystanders, and thought that some were still wearing their nightclothes. He could feel them glaring at him as he lit a cigarette, and before any of them could say anything he gave them a cheery wave, and hastily crossed the road to re-enter his showroom. It was 6 o'clock in the morning and he had been manoeuvring cars from around four-thirty that morning. The noise of the revving engines had woken them up and driven them out of bed to see what the awful racket was. Charlie was quite oblivious to the inconvenience and the irritation he had caused. His mind was made up: he had to get rid of this consignment as soon as possible. He was unhappy with the Packard, it was sluggish and repairs were expensive and many of the cars he sold required considerable mechanical back-up, so that the Packard was becoming a liability. What he needed to find was a motor engineer who could produce a car to his specification so that he could sell it exclusively with his own trademark. Charlie was undoubtedly a brilliant salesman, and had never had any qualms or conscience about using his title to gain entry to all the rich and powerful families in the land, or at least those to whom he believed he could sell a motor car. Not only was his record on the racing circuit a good sales tool, but also his prowess as an aeronautical balloonist, and on top of that Charlie Rolls could be unbelievably charming and entertaining when he chose to.

Fortunately none of his clients had seen him let fly at a reporter or policeman. Perhaps, had he managed to be slightly more polite and ingratiating, they may not have been so hard on him. As far as Charlie was concerned, the traffic police had declared war on him, and so he was merely expressing his own views to them regarding their unnecessary harassment. Typically, following a rather minor accident when he and Vera were returning from an event and were involved in a collision with a horse and trap where there were no major injuries, the reporter could not resist commentating that:

> The Hon Charles Rolls, who was thrown from his motor car the other day, and at the same time overturned a dog cart, injured one man's back and ankles, and gave another concussion of the brain, described the accident as "alarming, but not serious". One wonders what the Hon Charles would call a serious accident.

As Charlie felt that he was being deliberately persecuted by the police during every rally or tour in which he took part, he decided to get a cyclist to display a placard with the warning inscribed on it: "Beware – Police Trap just ahead". Eventually he was found out and threatened with prosecution, although at the time he was only doing 10 mph so the police could hardly pursue him on grounds of speeding.

One amusing aspect of these rallies was that the competitors gave names to their motor cars. Charlie applied the name of *Petrols* to his motor car, a nickname he had acquired at Cambridge. Some criticized him for entering too many tours and rallies, but, always with possible sales in mind, he ignored those comments. He only entered the race if it was relevant to whatever current project he happened to be involved in. Of course he enjoyed racing, but one of his favourite sayings was that "to win, it is necessary to finish". The trouble was, Charlie was becoming increasingly frustrated with cars where the engineering did not always live up to the demands of the challenges being given. The truth was, he was desperate to have a motor car built to his own specifications.

Claude Johnson, a friend and partner in Charlie's business and secretary to the Automobile Club, who was also a brilliant salesman, realized Charlie wanted to make radical changes in C.S. Rolls and Company. He discussed the possibilities of a change within the business with a friend called Henry Edmunds, who thought that it might be a good idea if Charlie were to meet a motor engineer called Henry Royce. He told Claude Johnson that, although he was "very dull and of a taciturn nature", he was nonetheless an extremely good engineer, and what he needed was someone like Charlie to take him on. Royce had a car to sell and Rolls wanted to replace the Packard. Charlie was certainly interested in the idea of exploring the possibility of forming a partnership with a good engineer so that he could ultimately have his own brand of motor car. When a visit to Manchester to meet Henry Royce was suggested by Henry Edmunds he said that he could not possibly go all the way to Manchester just then, he was far too busy. So, on 26th March, 1904, Edmunds wrote to Royce:

> I saw Mr Rolls yesterday, after telephoning you, and he said it would be much more convenient if you could see him in London, as he is very much occupied: and, further, that several other houses are now in negotiation with him, wishing to do the whole part of his work. What he is looking for is a good high class quality car to replace the Packard, preferably of three or four cylinders. He has some personal dislike to two-cylinder cars. I will do all I can to bring about this arrangement with Mr Rolls: for I think your car deserves well: and ought to take its place when it is once recognized by the public on its merit.

Edmunds then wrote to Charlie and sent him specifications of the Royce car adding:

> Knowing as I do the skill of Mr Royce as a practical mechanical engineer I feel one is very safe in taking up any work his firm may produce.'

These tactics might have failed. Royce did not want to leave his workshop in Manchester, and Charlie liked to get his own way. But Charlie could see from the specifications and sketches that this might be just what he was looking for. He sent a curt note to Edmunds on 29th April, asking him to come to Manchester with him the following week, on any day that was convenient, and that he intended to travel by train. A meeting was duly arranged. During the journey Edmunds recalls that Rolls had confided to him that he wanted a car connected with him that would in due course become a household name.

It was probably true to say that Charlie had more experience with almost every car produced than any other motorist in Britain. It took him no time at all to appreciate the obvious talent of Henry Royce, and he had, in his wide experience, come across many engineers and designers. Both men recognized their complementary strengths so that each of them could contribute that which the other lacked, but needed. Royce shared the same single-minded dedication to a project that Charlie had. Yet Royce was very Victorian in his outlook, and had absolutely no sense of humour. And despite his exceptional talent in engineering and design, he did not have the drive required to promote his work. Royce knew exactly how well connected Charlie was and that he could never hope to move in the social circles that the Honourable Sir Charles Rolls and his father moved in; nor indeed did Royce want to. In fact, he didn't want to move far from his factory, if the truth be known, and he wanted no involvement at all in the unsavoury task of marketing. The arrangement that Charlie and he ultimately made in naming their company "Rolls-Royce" suited Royce admirably. Royce knew perfectly well that without someone like the Honourable Charles Rolls, he would have finished his career in obscurity.

It took another nineteen months before The Honourable Sir Charles Rolls and Mr Henry Royce took the first steps towards amalgamation and it was not until March 1906 that Rolls-Royce Limited was born with a nominal capital of £60,000.

Although Charlie was the instigator of this extraordinary legacy, which still exists today, it should also be recognized that he was not just the entrepreneur, the visionary, and the marketing genius he proved to be, but in addition he had a flair for design. It was evident that he lacked the technical skills of committing accurate designs to paper, and usually conveyed his ideas in rough sketches. Most of these appeared to have been lost or destroyed, until some of his private papers came to light following the death of his sister, Lady Shelley-Rolls, when the Rolls' London home, South Lodges, was sold. Subsequent historians have been patronizing about Charles Rolls' abilities in the design field, but one should look carefully at his formative years, when no one could dispute his extraordinary inventiveness, and subsequently how he studied and understood the science and technology of all kinds of engineering.

CHAPTER SIX

Difficult challenges

Charlie Rolls now had to juggle his new company with his new passion, aeronautics, an expensive and unprofitable occupation but one which he believed would ultimately lead to a revolution in travel. It was clear that Rolls-Royce Limited was going to be used to fund aeronautical experiments, and in order to do that Charlie worked round the clock to make Rolls-Royce Limited a success not only England, but in Europe and the United States.

Charlie had the most extraordinary ability in judging talent, and just as he knew that Henry Royce possessed all the qualities he needed to launch his company, Rolls-Royce Limited, he also recognized Eustace and Oswald Short's dedication and unique skill that they applied in manufacturing their balloons and aeronautical experiments; he could tell that they were exceptionally gifted. Charlie had immediately recognized Eustace and Oswald's fanatical single-minded aims of developing their ambitions, although it appeared to him that they were overstretched and underfinanced. From his own experience he knew how it was all too easy to get carried away in the passion to find improvement, without calculating the actual cost. Many a time his father, Lord Llangattock, had put the brakes on. Nevertheless, he respected his father's careful accounting, for without it he and Royce could not have been the success they were at that time. Charlie appreciated his father's support and he appreciated his generosity when it came, sometimes at the least expected moment. He knew how fortunate he was, for he knew that many a brilliant venture had been ruined through lack of finance. It was no wonder that he despised many of his idle, rich friends, who had laughed patronizingly, and turned their backs when invited to invest in the motor car. It was not surprising that he sought the company of working men, rather than the aristocracy his mother expected him to associate with. Of course, that is not to say he despised all the upper classes. John Moore-Brabazon, Frank McClean and the young American he had met recently, Cecil Grace, were equally passionate about furthering the revolution of transport. Charlie Rolls regarded these men among his allies in breaking the boundaries of the twentieth century.

Charlie regarded Vera as one of his closest allies. Vera's love and support was something Charlie appreciated deeply, but somehow he did not know how to reciprocate. Although he was able to challenge convention on many fronts his restricted upbringing and starvation of female love when he was a child clearly had a terrible effect on him. He obviously adored and admired Vera more than any other woman and yet he hesitated when it came to breaking the social convention of marriage between families of different backgrounds. He may well have come round to it in the end as he was essentially always loyal and honest in his basic beliefs.

Vera was the stronger emotionally. She knew how she could damage Charlie, a girl from a trade background marrying an aristocrat – and not just any aristocrat. Vera more than anyone believed in Charlie's destiny. She had absolute faith in his goals, she was sure that he would achieve anything he set his mind to and she believed that her relationship with him must never come between his great ambitions. It was quite out of the question at that time for someone with a trade background to marry someone of the aristocracy. Frank also believed her decision to marry the son of a family friend, Iltid Nicholl, was the right one. Society would never have accepted Vera, and that would have angered Charlie so much that it could have had catastrophic effects on his career. If Vera was a respectable married woman no one would dare to criticize her friendship with Charlie. Fortunately, Iltid was aware that Vera loved Charlie but didn't care; he always hoped that she would come to love him in the end. He told Frank that he had enough love for them both.

Frank and Vera broke the news to Charlie together. The occasion was unimaginably painful for all three. Vera tried to reassure Charlie that she would always be there for him. Charlie was bewildered at his own helplessness in the face of social convention, which he abhorred. He was with Vera for his first motor car rally, with her for his first ascent, with her when she proposed the formation of the Aero Club, and she was the only woman he could bring to The Players' Theatre, the music hall, which they both loved. His solution was to throw himself into even greater challenges.

It was evident to Charlie that the Short brothers' technical knowledge of aeronautics was ahead of anyone he had met so far. Their workshop was a hive of activity, and at first glance may have seemed confusing, but then Charlie knew all about workshops, and he himself was in no position to criticize; without his long-suffering secretary he knew that his business would quickly descend into utter chaos. Eustace treated Charlie with a certain amount of diffidence as he pointed out the balloons in production, and the experimental prototypes they hoped to manufacture when they had the necessary finance. Clearly their order books were full, and Charlie calculated that at that time they must have had about eight men working for them. He wondered if they were securing deposits or payments in advance, but hardly liked to pry too deeply into their affairs.

Oswald was slightly taller than Charlie Rolls, and although he was still so young and appeared to Charlie to be a mere teenager, he spoke with the confidence of a mature man. It was disconcerting, and Charlie imagined that some clients would try to take advantage of the fact that he looked so youthful and would not take him seriously. He felt relieved that now most of the brothers' clients were friends of his and members of the Aero Club. Nevertheless he sympathized with the brothers' natural shyness but realized that it could be a disadvantage when promoting some of their ideas and designs. Charlie examined the intricate plans and drawings for future balloons on Oswald's drawing table, and listened carefully to the explanations on technical data. At the same time he could not help noticing alongside these sketches drawings of the Wright brothers' experiments taken from illustration in magazines and newspapers. He said nothing at the time but was impressed. Oswald was keen to Invest in the experimental construction of a copy of the Wrights' Flyer, and to a certain extent the increased business allowed this. Time was always an enemy, and there was never a minute to spare during the day. He drew a blank when trying to glean more information from America.

The British Government completely dismissed the whole importance of the Wrights' flight, so Wilbur and Orville lay low back in Dayton, Ohio. Meanwhile other pioneers in Europe were working independently to achieve a powered flight. Oswald continued to burn the midnight oil, reading a mass of aeronautical books and articles, and continued drawing crude but inventive experimental designs. The brothers allowed themselves very little leisure time, even though both of them were excellent sportsmen. They were both good footballers who played for their local clubs, and Eustace was a talented boxer. It seems that their sport did keep them fit enough to cope with the pressures of their physically and mentally demanding business. They understood how important physical fitness was. Unfortunately they were often a disappointment to their coaches, who would have liked them to attend more training, recognizing that both the boys were talented and an asset to their clubs, despite their obvious shyness and self-deprecating personalities.

Oswald told Charlie that it was their elder brother Horace who was the real genius of the family. Charlie found it hard to believe that their brother could be cleverer or more inventive than they were. But Oswald assured Charlie that Horace's skills as an engineer were ahead of his time. Without Horace, Oswald told Charlie, Sir Charles Parsons could not have developed the turbine engine in its present form. Horace was a strong character, he said, and had a sense of humour that some people found outrageous, and he was not sure that he understood it himself. He was also quite a showman, and when the *Turbinia*, which he had helped design and build for Sir Charles Parsons, was to be displayed at the Spithead Royal Revue, he suspected that he would not get the attention he deserved. So he decided that, prior to the actual review, in which the Navy's crack destroyer was briefed to steam past a line of ships at anchor, ostensibly to clear the way for the Royal Yacht, but actually to show off her speed, Horace would set off the little *Turbinia*, which shot out of the line of ships and left the destroyer standing. They were furious with him for ruining the display but he had proved his point. In addition, the invention of the Auxetophone was Horace's, not Sir Charles's, but fortunately Horace had had the good sense to patent it.

Oswald told Charlie that unfortunately Horace had never shown any enthusiasm for their balloon business, but since the Wright brothers' launching of their heavier-than-air machine, Horace had become convinced that that was the way forward, and sooner rather than later aeroplanes would take to the air as a permanent form of transport.

Charlie wanted to meet this brother for himself.

Meanwhile Colonel Templer had retired as Superintendent of the Army Balloon Factory and Balloon School, and was succeeded by a Colonel Capper, who was also a keen balloonist, kite and airship man, and had accompanied Charlie and Brab in the *Britannia* in the Gordon Bennett Balloon Race. On Templer's recommendation Capper went to the USA to study the development of aeronautics, and during his visit went to Dayton to meet the Wright brothers. They formed a very good relationship, and subsequently wrote to each other to keep abreast of activities on either side of the Atlantic. The Wright brothers trusted Capper and in 1905, through Capper, they offered rights to their invention through the War Office to the British Government. The whole thing was handled with appalling ineptitude, ignorance and lack of imagination and no one in His Majesty's Government even bothered to investigate the huge prize they were being offered.

After a long period of procrastination the whole deal collapsed, and the opportunity to seize the initiative in powered aviation in Europe was lost. Some of the blame could be attributed to the Wright brothers themselves for their determination not to give away their invention too cheaply, although they could hardly be blamed for that. Some of the blame had to be attributed to Capper, who had become interested in developing an aeroplane by the British Army itself, using the designs made by a young Army Lieutenant, John W. Dunne, under the umbrella of secrecy in Blair Atholl in the Scottish Highlands. Ultimately this venture failed, and the Government, despite the warnings of success of achievement coming to fruition in Europe, turned their back on the possibility of powered aviation, but Lieutenant Dunne would still play a part in the birth of aviation.

Oswald wished he could talk to Horace, and when the Aero Club, which had now come to rely on the Short brothers to produce most of their balloons, made the brothers the official engineers of the Aero Club, Oswald felt it was time to write to his brother and tell him about their recent achievements, and their ambition to emulate the Wright brothers' heavier-than-air machine. He also pointed out the Government's total lack of interest in aeronautics, the danger that Britain faced by ignoring what the rest of the world knew sooner or later would become a reality, and that they would be completely left behind because of their intractable attitude.

When Horace received "The Kid's" letter he realized he should find out exactly what Eustace and Oswald were up to. He needed to pay them a visit in due course and bring his family with him. His family now included three children: Leonard, born in December 1900, Ruby, born in 1903 and the baby Francisco, usually shortened to Francis, born in 1905. His mother, of course, could hardly wait to see her grandchildren.

CHAPTER SEVEN

Leysdown

The small and elite members of the Aero Club met regularly and shared their experiences together and learnt from each other. All the members were impressed with the Short brothers' diligence and enthusiasm, and having appointed them as their official aeronautical engineers, they had regularly commissioned them to build or recondition balloons. Griffith Brewer, an early member of the Aero Club, and John Moore-Brabazon were frequent pilots in the Gordon Bennett races which took place in France and America, and Eustace and Oswald were all too aware that their reputation was on the line for these events. Gradually the members became more ambitious, and as the Short brothers began to experiment on their own account, they were running out of space in the Battersea workshop, and were desperate to find a more private workshop away from the public and the press. For all their confidence in their work and their success with members of the Aero Club, Eustace and Oswald struggled to keep on top of their cash flow.

The Aero Club were acutely aware of the Government's refusal to entertain the development of the manufacture of aeroplanes. In contrast, the French, Germans and Russians had realized the significance of the development of the aeronautical experiments their own pioneers were developing and had begun to invest in their research. Not only were the British Government hostile to any investment but the press, particularly *The Times*, supported their view and wrote damaging and critical articles concerning the pioneers so that the general public were also of the view that the pioneers were a lot of crackpot experimentalists. It seemed that the only newspaper that supported them was the *Daily Mail*, owned by Lord Northcliffe, and the only journalist who wrote encouraging articles and followed the development with enthusiasm was the *Mail*'s Charlie Hands.

As a result of these circumstances, in 1907 Frank McClean called a meeting with members of the Aero Club and decided that they needed to find a place well away from the public view where not only could the Shorts set up new workshops but so too could the Aero Club; and they also needed private practice flying grounds for their experiments. Griffith Brewer volunteered to find a suitable place, and as it happened he had already formed a fairly good idea of the perfect location following an ascent he had made over Kent earlier that year.

When Griffith Brewer crossed the River Swale on the King's Ferry, the bridge that connects the Isle of Sheppey to the mainland of Kent, and the mist that had followed him over the bridge disappeared in flurries across the marshes, a strange feeling of destiny enveloped him. He stopped to allow some wandering sheep to cross the road, and watched the marsh birds swoop and call to each other in the distance. He drove slowly on, turning right at the T-junction following the signs to Eastchurch and Leysdown, which he had identified from an Ordnance Survey map as having a golf course, which he thought might prove flat enough for flying experiments. He decided to stop for an early lunch and found a pub next door to the church in Eastchurch called The Castle. He was surprised to see a Gordon Bennett motor car parked outside; he drew up behind it and parked his own brand new Rolls-Royce motor car. There were a few local villagers propping up the bar; it was obviously lunchtime and they were having a pint of the local beer with a lump of cheese and a hunk of bread. As he walked into the bar the conversation died momentarily, until one of the men declared cheerily, "You ain't from these parts then?" When they discovered it was not only Griffith's first visit to Eastchurch but also to the Isle of Sheppey, it did not take them very long to expound on the local history of the island.

"Julius Caesar was our first invader," they began, and by the time they had gone through the various wars with the Danes and reached the invasion of the Saxons, their boss had arrived.

Charlie Rosewarne was the local builder in Eastchurch, and he was also the owner of the Gordon Bennett.

"Go on you lot, back to work, this ain't a 'oliday y'know!" said Rosewarne cheerily.

"We was just tellin' this gentlemen abart all the ghosts and that on the island. He's a visitor. We was just fillin' 'im in. Maybe 'e'd like to buy one of your 'ouses, Charlie!" said the one who had started to tell Griffith about Julius Caesar and the Danes.

"Go on, off you go, I'll see you later," said Rosewarne and the men left laughing and joking amongst themselves.

Charlie Rosewarne held out his hand, dropped the Kent accent and greeted Griffith. "How do you do, my name is Charles Rosewarne and I'm the local builder, coroner and head of the darts team in Eastchurch," he said with a laugh.

"How do you do," responded Griffith. "My name is Griffith Brewer, I'm from London and I was on my way down to Leysdown where I believe there is a nine-hole golf course and a house nearby."

"That's right," said Charlie. "It's owned by a Mr Andrews. If you wanted to rent it, I'm sure he would be amenable."

"I'm just looking around at the moment," replied Griffith, not willing at this point in time to divulge the purpose of his visit.

"Here's my card," said Charlie. "If you want any help finding your way around or you want to know anything, call on me. I've lived here all my life; there's not much I don't know about Sheppey. It's a small island and it certainly has a strange history. Just this small village alone has a host of ghosts from the past. But I won't keep you, I expect you want to get on."

"No," replied Griffith, "I've got plenty of time and I'd like to hear some of the history."

"It would take weeks to tell you about all the history of the Isle of Sheppey, but I'll tell you a bit about Eastchurch, I can't stop long, mind; I need to see what my layabouts are up to!"

"Of course, I quite understand," replied Griffith.

"Anyway," began Rosewarne, "as you leave Eastchurch for Leysdown, you will see Shurland Hall on your left set up on a small hill. We're working there now, as a matter of fact. It was originally owned by Sir Robert de Shurland, Lord of Shurland and Minster during the reign of Edward the First in the thirteenth century. He was a great favourite with the King but had been by all accounts a rather unattractive man, small, bandy legged with powerful shoulders and a bull neck, exceptionally strong, and greatly feared by his enemies. He was created knight baronet for his bravery at the famous siege of Caerlaverock. During one of the many battles in which he became involved, he killed a monk who dared to disobey him, in a fit of rage. Fearing the consequences of this terrible sacrilegious crime, which was almost unpardonable, Sir Robert decided to try to see the King in person to beg for mercy. By chance the King had been held up by bad weather just off the coast, near Minster. Realizing that immediate action was called for to save his reputation and possibly his life, Sir Robert saddled up his magnificent charger, Grey Dolphin, and with him swam out to the King's warship to beg for forgiveness. The King was so impressed by Sir Robert's explanation and apology that he pardoned him on the spot. Sir Robert plunged back into the sea and climbed back onto Grey Dolphin's back to swim back the way they had come. As he emerged from the sea onto the shore with Grey Dolphin, an old hag appeared out of the mist and gloom and accosted him. She taunted him over the crime he had committed and prophesied that his horse, who had so gallantly saved his life that day, would within a year be responsible for his death.

"Sir Robert, elated with the result of his mission but furious with the old hag's prophecy, leapt off Grey Dolphin, drew his sword and with one murderous blow severed the head of his faithful charger to prove the impossibility of such a prediction. As he buried him on the shore by Scraps Gate, the hag faded away into the gloom. Many months later, walking along the shore deep in thought he kicked a protruding object in the sand, violently, and as he proceeded along the shore he came face to face with the old hag who had prophesied his doom. Grinning hideously she told him that he had just kicked the skull of Grey Dolphin and disappeared, cackling, into a mist that had suddenly swept across the shoreline. Sir Robert could see that his boot had been pierced and his foot was bleeding badly. During the following weeks the wound would not heal, poison set in and shortly afterwards he died.

"You can see Sir Robert's full-life-size effigy, resting on his shield with the life-sized figure of a horse's head emerging from the sea, lying in the Abbey Church of St Sexburga, Minster on Sea.

"After two hundred years, Shurland Hall passed from the de

Shurlands into the Cheyne family, when Sir Robert's only heir, his daughter Margaret de Shurland married William Cheyne. It remained in the Cheyne family until the reign of Elizabeth I and was reputed to be one of her favourite country retreats. Henry VIII often stayed there and he and Anne Boleyn spent part of their honeymoon there.

"So you see, Mr Brewer, that's just a taster."

"Fascinating," said Griffith. "So tell me, who lives there now?"

"The Cunninghams. Mr Cunningham is a banker; he's away during the week. Mrs Cunningham is there most of the time. She's a pianist, plays the organ in our church. She is a good friend of my wife, Kate, and her daughter Florence is a friend of my daughter's. Of course most of the locals won't go near the place, terrified of coming face to face with Sir Robert de Shurland, or Anne Boleyn with 'er 'ead cut off." Charlie roared with laughter. "Well, if you'll excuse me, I must be on my way. Contact me any time on that address on my card if you'd like another history lesson," laughed Rosewarne. "Ta ta for now!"

"Delighted to meet you, Mr Rosewarne; thank you for telling me the story of Shurland Hall," called Griffith as Rosewarne left the pub with a wave to everybody in general.

Griffith Brewer was very pleased to have made the contact with the builder, Charlie Rosewarne, realizing that he might be very useful in the future. Rosewarne was obviously a popular member of the community.

Leysdown was no more than a fifteen-minute drive from Eastchurch through rolling rural countryside, on narrow roads that were little more than farm tracks. The sign post to the "Golf Course" brought him to Muswell (Mussel) Manor, which was certainly an isolated spot. The small sixteenth-century manor house stood surrounded by flat, undulating land without another property in sight, and the sea was only a few hundred yards away across two fields where the waves lapped gently on to a shingle beach divided by breakwaters. The nine-hole golf course stretched out to one side of the property but the course was clearly used very spasmodically because the greens and fairways were only attended when the groundsman had time from his other rural activities. The property appeared to be uninhabited, apart from a caretaker who seemed to come and go, but Griffith felt it would probably be useful to maintain her employment. It did not take Griffith long to make the decision to obtain the agent's address. It seemed perfect for the Aero Club's experiments, and he returned to London that evening to report to Frank McClean that he had found an ideal spot where the Aero Club could operate in relative privacy. As Rosewarne had said, the property was owned by a Mr Andrews who was also a builder. Griffith arranged a meeting with Mr Andrews, Charlie Rolls, John Moore-Brabazon, Frank McClean, Frank Hedges Butler and Eustace and Oswald Short at Mussel Manor to discuss terms to rent the land and property from him.

Eustace and Oswald and the other members of the Aero Club were delighted with Griffith's find, and Frank McClean immediately leased land and Mussel Manor from Mr Andrews. Mr Andrews, of course, was also absolutely delighted: he could see the potential, as a builder, as to how profitable this venture was going to turn out, and left the meeting with the members of the Aero Club a very happy man with the prospect of a substantial business deal.

Left on their own, the members got down to the practicalities of the move, and ended by discussing the terrible apathy of the Government and the way the newspapers apparently saw them as the idle rich wasting their time. Far from being despondent, they were more determined than ever to prove the Government, the public and the newspapers wrong.

Charlie Hands of the *Daily Mail*, the lone voice of disagreement with the Government and the other newspaper journalists, did not waver in his belief that the pioneers would succeed in their ultimate ambitions. Hands knew that young men like Charlie Rolls had already proved that the motor car was here to stay. He had flown in the face of public opinion, and now all those who had mocked him at the beginning of the century were eating their words. It seemed obvious to him that Charlie Rolls was not the sort of person to waste his time on something that was not important: he was far too parsimonious to do anything like that. He desperately needed good press to sell his motor cars, but on the other hand, given the terrible press that he had had over the years it was not surprising that he viewed the press with deep suspicion, and was often exceptionally rude. Charlie Hands was nevertheless determined to stick with him, because whatever one thought of Charlie Rolls, no one could deny that he was daring and fearless.

Charlie Hands had also tried to interview the two Short brothers, who he knew were the Aero Club's exclusive balloon makers and engineers, but they refused to cooperate in any way. He could not understand why that was, for after all, publicity would be good for them. Clearly they did not think like that. It seemed to Charlie Hands that they would go to any lengths to avoid speaking to him. He decided to bide his time and respect their privacy. Other members of the Aero Club were much more forthcoming. John Moore-Brabazon, for instance, was a particularly open sort of character, full of energy and good humour at practically all times. Frank McClean and Frank Hedges Butler were also good humoured and generous, approachable, characters. But at this time none of the members viewed the press to be on their side, and they were even guarded with Hands.

CHAPTER EIGHT

Charlie goes to the USA

In 1907, while Eustace and Oswald began to make arrangements to set up additional workshops in Leysdown and the Aero Club were contracting Mr Andrews to build them their own workshops, Charlie was crating up his brand new Silver Ghost Rolls-Royce for the New York Motor Show. Naturally he wanted Vera and Frank to come with him, but that simply was not possible. Vera had been married for barely a year and although Vera made every argument she could to accompany Charlie, Frank's common sense prevailed. The gossip would have damaged them all. Charlie asked Griffith Brewer instead, explaining that they would use the excuse of the trip to go to Dayton, Ohio to see the Wrights.

Charlie and Griffith boarded the RMS *Lusitania* with the precious Silver Ghost Rolls-Royce in her hold in September 1907. It was her maiden voyage and naturally the publicity was immense. Once again Charlie had achieved a PR coup in announcing his imminent arrival in New York. Not only did the *New York Times* give him an extravagant spread in advance with a very dashing photograph of himself beside his Silver Ghost, but he managed to acquire a free passage by using his title and persuading the owners of the liner that his presence on the maiden voyage would be a superb advertisement for the *Lusitania*, pointing out how fortunate they were to have him on board. Griffith had to work hard the whole of the crossing to make sure that Charlie portrayed himself as charm personified. It was not an easy task.

When they arrived in New York, Charlie and Griffith were whisked off to the Waldorf, courtesy of the Rothschilds, and given a magnificent suite from where they attended the motor show. The matrons of New York could hardly contain themselves with delight, vying to meet this handsome English aristocrat, so charming, so eligible. Charlie did brilliantly, sales exceeded all expectations despite the 1907 stock market crash, but he could barely wait to leave the show to go to see Wilbur and Orville Wright. He felt Royce could hardly complain that he was wasting Rolls-Royce Limited time after his extraordinary success in New York, not to mention a free passage and free accommodation.

Charlie and Griffith boarded the train at the Grand Central Station and sighed with relief as they sank into their seats on the Great Western Railroad for Dayton, Ohio. They had telegraphed Wilbur and Orville in advance so that when they arrived in Dayton, Wilbur and Orville were there at the railroad station to meet them.

Charlie and Griffith were able to tell the Wrights first hand of their own difficulties with an intractable government and a hostile press. It seemed that their problems were identical. Charlie's obvious enthusiasm despite the setbacks was a tonic to Wilbur, who was easily depressed by circumstances. Charlie, in a fit of optimism, offered his services as an agent and furthermore offered to be their test pilot, suggesting that they crate up one of their fliers and send it to England. It was a very extravagant and brave gesture considering that a) he had never been in an aeroplane before, let alone flown one and b) how on earth did he think he was going to have time to take on such an overwhelming task! Fortunately, Griffith quietly stepped in and offered his services as their agent in England. Wilbur and Orville were delighted to accept the offer.

Reading between the lines, Charlie Rolls' approach to Orville and Wilbur to become their agent was perhaps too intense. Although utterly genuine in his offer to be a test pilot, he probably appeared to be too ambitious, too direct. Nothing was impossible as far as Charlie was concerned, no one could thwart him, but they must have been aware of his inexperience. One can imagine that he would have looked deeply, directly and unnervingly into Wilbur's face, searching with those large, glittery, brown eyes for acknowledgement of the proposals put to him. It was probably part of the reason that Orville and Wilbur chose Griffith Brewer as their agent in the end. He was a gentler, steadier and more approachable person who was nevertheless close to Charlie Rolls and an important member of the Aero Club. Perhaps they also felt that Charlie would find it difficult to find the time to act for them when he was so involved launching his new motor car company. Rolls-Royce Limited was already beginning to make a prestigious name for itself in the United States as well as in England. They must have felt that Griffith Brewer would have more time, and his sensitive and diplomatic personality would be more in tune with their own approach. Clearly Charlie was perfectly happy with this arrangement, and in fact he came to rely on Griffith for advice and help and companionship, and trusted him implicitly. Griffith Brewer was a gifted engineer himself, who had begun ballooning in 1891 with his wife.

Having boosted each other's confidence, Charlie and Griffith set off back for England in an optimistic frame of mind while the Wrights, equally optimistic after Charlie and Griffith's visit, set out to approach governments on their own behalf.

Once again the Wrights approached the American Government but still the officials failed to take them seriously and were almost patronizing in their refusal to take any interest. They wrote to the Military Attaché in Washington and were stung with the reply which came back, saying: "I regret to inform you that the Army Council has decided, after most careful consideration that, especially in view of the great cost, it would not be advisable to purchase your invention."

Wilbur and Orville then turned their attention to the British Government. There they were met with equal scepticism, and indeed rudeness, by the Minister of War, Richard Haldane, who replied curtly to the Wrights' incredible offer of fifty aeroplanes for £2,000 each that "the War Office was not disposed to enter into relations with any manufacturer of aeroplanes".

A similar response was sent to the Wrights by Lord Tweedsmuir, First Lord of the Admiralty: "I have consulted my expert advisers with regard to your suggestion as to the employment of aeroplanes and I regret to have to tell you, after consideration of the Board, that the Admiralty, whilst thanking you for bringing the proposals to their notice, are of the opinion that they would not be of any practical use to the Naval Services."

Meanwhile the Wrights visited Europe and tried to make progress with the French, who by this time had decided that the brothers were not capable of the flights they had claimed to have made. After all, they had not flown for two years, and Henri Farman had apparently made his first circular flight, according to *The Times*. In addition the French had their own Louis Blériot and the Voisin brothers who as far as they were concerned were way ahead of the Wrights. They were met with equal disdain by the Germans and the Russians. Disgusted by the way the British Government had dismissed the Wrights, Charlie Rolls and Griffith Brewer decided to keep in close contact with them, convinced that the Wrights were streets ahead of the French or Germans.

Wilbur and Orville knew it was time for action, and that they were going to have to demonstrate their Wright Flyer themselves. They returned to Dayton, Ohio and crated up a Wright Flyer at the end of 1907, and shipped it to France ready for demonstrations in 1908. They set up camp at Kitty Hawk and learnt the technicalities of flying their aeroplane all over again. Hours and hours of practice and modifications finally achieved the desired result and the confidence that they could demonstrate their flier without fear of an accident or mishap. They prepared to return to Europe the following year, confident that this time they would not fail to impress the public, and that they would finally get the attention of governments.

Shortly after Charlie and Griffith returned to England, Frank McClean called a meeting at Mussel Manor, which Charlie, Griffith and Brab attended and of course Eustace and Oswald, who still worked out of their workshops in Battersea as well as their new premises in Leysdown. That morning they were catching the "Sheppey Light" from Victoria Station, and there was the usual noise and hubbub of activity, porters pushing their long wooden barrows laden with luggage, calling above the noise of the steam engines to "mind your backs there!" Everyone shouted their conversations to be heard over the noise of the firemen shovelling coal into the engines, as the engines spurted billows of steam, the drivers tested their horns and the constant arrival and departure of the trains themselves added to the general chaos. Vendors shouted their wares of fresh fruit and flowers, while newspaper sellers called out their papers' headlines. The noise was almost deafening.

It was with the greatest relief that Oswald and Eustace finally pushed through the crowds to reach their platform, climbed into a carriage, and shut the door with a clang which could barely be heard against all the other noise. Eustace found the sports page in *The Times* and settled down to read the football news of clubs around London. Oswald opened a book he had borrowed from the library on Darwin. It was not long before they heard the engine's piercing whistle signalling its departure. Suddenly the door to their carriage was flung open and a man propelled himself into the middle of it, landing on all fours on the floor between Oswald and Eustace. A guard closed the carriage door with a bang and gave a piercing blast on his whistle as the man clambered to his feet. He dusted his heavy green tweed suit down, clutching the luggage rack above for support as the train lurched forward. He retrieved his hat and brown leather portmanteau from the floor.

"I do apologize, for intruding on you like this. My name is Charlie Hands of the *Daily Mail*."

Oswald put his book down, horrified that he and Eustace had been followed, while Eustace doggedly drew the copy of the sports page closer in front of himself.

"I have never been able to speak to you before and I really wanted to have an opportunity to tell you face to face that I am on your side. You do have Lord Northcliffe's support, my boss, and I personally have followed and supported Mr Charles Rolls around the country for several years, attending the Gordon Bennett races, watching his interests in ballooning gradually turning to the heavier-than-air machines. I can see Mr Rolls has a fearless pioneering spirit and I have faith in his objectives, as I know you yourselves do. I would very much like you to trust me. Eventually this Government will have to pay attention."

Eustace lowered his newspaper. Oswald held out his hand, which Charlie shook. "Oswald Short," he said; "my brother Eustace," and Eustace held out his hand, which Charlie also shook.

The ice was broken. Charlie Hands was a sincere and warm-hearted man and even Oswald and Eustace could see that he meant well by his unorthodox introduction. Oswald knew that they needed support, but events were escalating so fast and changing at such a rate that what was discovered one day in the workshop was nearly always overtaken by something new the next. Precision and accuracy were the name of the game, and if they faltered they knew it would cost the lives of those who were putting everything at risk and who seemed to know no fear. Oswald was keenly aware what a frightening responsibility it was to hold a man's life in the safe keeping of his machines. How could he live with his conscience if his machines were the cause of an accident and perhaps an injury or death. There was many a time Oswald awoke sweating and afraid in the night.

The Sheppey Light expelled a fountain of steam as it came to a halt on the tiny platform in Eastchurch, which was no longer than the train itself. The stationmaster, doubling as the guard and ticket collector, was decked out in a resplendent uniform. He strutted down the platform bowing obsequiously to the alighting passengers, and welcoming them on to the platform as if it were hallowed ground. Eustace and Oswald alighted. Charlie had said what he wanted and remained on the train to return to London. The stationmaster slammed the doors, blew the whistle and the train began to trundle back to Queenborough. He ushered the two young men through the entrance with such ceremony that they were mildly embarrassed. Emerging from the station they saw a pony and trap waiting to take them to Leysdown for their meeting with Frank McClean at Mussel Manor.

CHAPTER NINE

The Mussel Manor Meeting

The interior of Mussel Manor was dark and gloomy even on a sunny day. The Aero Club members with Eustace and Oswald sat around a stalwart oak table underneath the dark beams. Flickering oil lamps cast eerie shadows across the mustard-coloured walls. Frank McClean sat at the head of the table.

"I felt that I must call you here today to warn you all of the gathering hostility and blindness of this Government towards our movement. It has been drawn to my attention that the German, French and Russian governments have now increased their investments with their pioneers to the equivalent of £400,000 per annum each."

The Times, he told them, had published reports from the War Office that said:

> Although in the existing state of aerial navigation Great Britain is not exposed to any serious danger by land, it would be improvident and possibly dangerous to assume that the rapid development which the art of aerostatics has recently made may not entail, in the future, risks by land and sea. Invasion of Great Britain in air ships on a large scale may be dismissed as unlikely for many years to come . . .
>
> The military authorities point to the fact that nowhere has any machine designed for flight in the air proved effective. All this points to a lack of practical working in the various designs. When it is possible to cross the Channel with a party of excursionists to land at any fixed point, the War Office may be persuaded to take recent experiments seriously.
>
> The Committee recommended a sum of £35,000 in the Naval

> Estimates for a rigid airship and £10,000 in the Army estimates for continued experiments on navigable balloons and . . . the experiments at military ballooning establishments with airplanes should be discontinued but advantage should be taken of private enterprise in this form of aviation."

"I repeat, gentlemen,' said Frank McClean, gravely, '". . . advantage should be taken of private enterprise in this form of aviation."'

The mood shifted to something like outrage but silence continued to pervade the room. Oswald swept his gaze around the table and wondered how many would sustain the enthusiasm, how many would find the funds, and who among them would retain the nerve to carry on. Mr Charles Rolls? A man driven by conviction and imagination, certainly brave and creative, but a man with a complex character; nothing would distract him. No, nor Mr Frank McClean. A nicer man you could not wish to meet; generous, warm hearted and sensitive to everyone's feelings and, Oswald felt, one who would always stand by you to finish what had begun. Following in the steps of his illustrious father, he too had become an astronomer. His inheritance from his father had been enormous but he had also made his own money. He was Horace's age, and Oswald was sure that Horace would be unable to bully him, for he was too nice to be bullied and in any case people looked to him for strength and encouragement. Mr Moore-Brabazon, a young man who was bullish and always wished to prove a point; he would be determined to prove himself. Mr Griffith Brewer, enigmatic, stalwart, and loyal to his friends, but quietly determined and single-minded. He was a good engineer in his own right and understood all the technical complications.

Then there was his own brother Eustace, who remained as obsessive as ever about travel by air, and was entirely dedicated to his own belief that his and Oswald's attention to excellence in detail and delivery of goods devoid of fault would achieve everything the flying pioneers wanted. He believed in his own talent, he believed in Eustace's and, yes, especially Horace's. Oswald could see that Eustace was, on the whole, uninterested in these reports from *The Times*; he was only interested in the progress with his men in the new workshops and wanted to leave the meeting now that the main business was done, and get back to work. The challenge: "When it is possible to cross the Channel . . . to land on a fixed point . . ." rang in his ears, drowning everything else out. He too wanted to get back to his workshops. There were problems needing attention, and a glider belonging to Cecil Grace had come down on the other side of Frank McClean's nine-hole golf course and needed towing in.

Oswald closed his eyes, studying these men's characters in his own mind. No, these men would stay until the bitter end; they were in too deeply, just as he and Eustace were, and he wished that Horace would grasp the scale of the challenge before them. Oswald knew that nothing would stop Horace once he had set his mind on something. His invention of the Auxetophone and his work on the turbine engine for Sir Charles Parsons had taken his entire life over at the time of their development, and he was sure that, sooner rather than later, the aeroplane could do the same. If only Horace and Eustace would get on better. He could tell that Horace was proud of their achievements and secretly thrilled that they had been nominated as the official Aero Club engineers.

The men remained silent and still in their seats, their hands placed lightly on the table in front of them, regarding their notebooks and the agenda Frank McClean had drawn up to put into production. They digested the enormity of the Government's intractable attitude. The derision heaped on the aeronautical pioneers from friends and the general public was one thing, but the plain stupidity of men in positions of power was actually frightening.

Moore-Brabazon broke the silence and slammed his fist on the table. "Not a scrap of encouragement as usual, not even a smidgen of curiosity, not a doubt in their blank minds that maybe, just maybe the French, Germans and Russians could be on to something. What's the sum the Germans are spending on aviation?"

"Around the equivalent of four hundred thousand pounds per annum," repeated Frank McClean quietly.

"So! Where does that leave us, pray tell? Well I'll tell you! In penury! That's where it leaves us! What did that letter in the *Observer* report? '. . . if pigs could only fly then we would all happily follow suit.' Well, I'll tell you! Yes! I know exactly what I'm going to do! I shall go to France!

"No Frank! We're wasting our time at the moment, not a soul gives tuppence halfpenny. The attitude here is absolutely exasperating beyond belief. Everyone is so ready to discourage one, ridicule one and look upon one as an amiable lunatic. Yes, I shall go to France where at least one is taken more seriously."

"Brab!" exclaimed Griffith Brewer curiously. "You're not going to leave us, surely!"

"No! Of course not, old boy. But I'm going to France to learn to fly, and I shall learn as much as I can in France, pick their brains and see what useful things I can bring back to the Aero Club. I am sure there will be plenty to exchange. Their Government is only marginally more informed than ours. It's the public over there that are so supportive. Nobody regards Gabriel Voisin as a lunatic, and most believe that Blériot is a hero, and if we're not very careful they are going to steal the lead on us. One could be forgiven, could one not, in removing our whole enterprise to France where at least one would be taken seriously and I don't mind telling you if it were not for Oswald and Eustace, and their brilliant workmanship, I would, and damn the lot of them!"

Oswald breathed a very long inward sigh of relief as he mentally totted up the unfinished orders he might have had on his hands.

"Look at society! They are totally derisive, they care more about their ridiculous house parties. Even Lord Montagu, who I had hoped had a modicum of intellect, announced to the House of Lords that 'the world of aeroplanes is a useful plaything of the wealthy' and 'the lunatic fringes of the engineering world'."

Eustace drew in his breath sharply and looked hard at Oswald, who returned his stare. The statement hit the brothers harder than the reports in *The Times*. "Lunatic fringes!" What did these people know about engineering? Oswald could see the colour rising from Eustace's neck and recognized the signs of deep anger. Were they up against brick walls? Then what did it matter, thought Oswald; he and Eustace knew all about brick walls, but that had never stopped them before. Frank McClean intercepted the brothers' silent communication and believed in that moment that the Aero Club were about to lose the finest engineers they were ever likely to get.

He was mistaken, of course, for he was unaware of most of the brothers' background. Frank knew that they had started their ballooning business when Oswald was a boy of fourteen, but he was unaware that at twelve he had left school and worked as a delivery boy just to help feed the family. It took more than an uninformed comment to stop the brothers in their tracks and in fact it served as an incentive to work harder than ever.

"Come, Mr Short," he said, addressing Eustace gently, but including Oswald. "Under no circumstances must you be put off by these public statements. They are indeed disappointing, but they are made by ignorant men without vision. Is that not so, Charlie?"

"Indeed!" responded Charlie, looking grim with a faraway look in his eyes. A cold fury took hold of him. He would deal with the contemptible, hypocritical Montagu himself, preferably in front of as many of his "chums" as possible.

"We must not lose sight of the fact that flight has been achieved, and like the motor car only some fifteen years ago, we have to prove its extraordinary possibilities for the future of the world." Charlie smiled thinly and continued, "It's too late now, we're almost ready to demonstrate this revolution." He turned towards Oswald and stared unblinkingly at him. "Don't you think, Mr Short?"

But it was Eustace who cut in and replied, "Sooner rather than later, Sir!"

"That's the ticket!" said Frank McClean, smiling warmly and encouragingly at Eustace and Oswald. "Now, it's a beautiful day, let's all cheer ourselves up and go and look at the splendid progress you are making here."

Oswald returned his smile and in his soft, unassuming voice reassured the meeting: "If we'd listened to criticism over the years, Eustace and I would be coal merchants now!"

Brab led the laughter, little knowing just how accurate that was. Oswald and Eustace looked faintly surprised because they always told the truth, and saw nothing amusing in that statement.

As they rose from the table, Charlie's mood altered; he threw off the cloud of gloom that had settled over him during their meeting and turned to Eustace and Oswald with a rare smile, which revealed his admiration of the brothers.

"I will prove to you both just how wrong these people are!" Turning to Griffith Brewer he said, "We must make sure that when Wilbur Wright returns to France we tell him about the Short brothers and endeavour to arrange a meeting. They must see for themselves the progress that has been made."

"We would very much appreciate a meeting," said Oswald quietly.

"We shall see what we can do," replied Charlie. "It is most important that you meet. In fact, we must all go", Charlie continued. "I believe I should ask Frank Hedges Butler and . . . well, of course." At that point he took his pocket watch out of his waistcoat pocket with his right hand and held it away from him, looking at it unseeingly; ". . . and perhaps his daughter Vera may care to accompany her father." Charlie coughed and went slightly pink, finding the pocket watch of the utmost interest.

Frank retrieved his pipe from the ashtray on the table and jammed it into his mouth to prevent a large grin. He winked at Griffith Brewer, who quickly turned away to hide a smile.

"Excellent!" cried Frank. We shall all look forward to the visit enormously."

Eustace and Oswald exchanged glances in silent communication. Neither of the brothers could conceal the surge of excitement they felt, and they quickly led the way out of the house, Oswald ducking under the low beams and the heavy oak front door frame. Oswald at twenty-five years old still looked like a schoolboy, handsome and aristocratic looking, despite his humble background. Like Eustace, who was also a good-looking man, but smaller, with a clear skin and fine bone structure, he was quiet and softly spoken, with faint traces of his Durham accent. It was natural for Oswald and Eustace to lead the way to their workshops. They were not subservient in any way to these rich aristocrats, and they in their turn respected the brothers as equal pioneers.

It was a perfect summer's day. The sea sparkled, the sand shone like pale gold, and the green fields and little golf course running down to the shore invited you to enjoy their beauty. Gulls and herons swooped and dived across the wild, empty landscape.

Charlie, momentarily blinded by the light, saw the shadows of the birds against the sky. "Flying silver ghosts," he muttered under his breath.

"Do you know," announced Frank McClean cheerily, "I'm going to have nine holes and a dip later! Who is going to join me? Griffith? Brab? Charlie?"

Oswald and Eustace hurried ahead across the intervening field to their workshops, the familiar sounds of hammering and sawing drowning out the lap of the waves on the beach. The workshops were twice as big as Battersea, big enough to store at least four aeroplanes. There were four constructions, made mainly of steel, with high, sloping corrugated iron roofs strong enough to withstand the winter gales. Each flying shed was able to hold a workforce of up to fifteen men. Eustace and Oswald had recruited their men from the best joinery shops around Battersea and the East End of London. The brothers disappeared inside. They quickly took off their jackets and retrieved their long white carpenters' aprons and joined their men at the work benches.

Frank, Charlie, Brab and Griffith stopped halfway across the field and fell into discussion once again. The sea breeze ruffled their hair, for they had all left their hats inside Mussel Manor. Frank turned his face to the sun and breathed in the salty air. As he did so he caught a faint and unusual sound in the distance.

"Hear that?" he asked the others, turning his ear in the direction of what was now an unmistakable sound of horses' hooves accompanied by unmistakable cries of "Yeee ha!" and "Yippee!"

The men used their hands to shade their eyes against the sun to see who on earth was riding at such breakneck speed along the sand, dangerously jumping the timber, low over the necks of their mounts, while the horses' manes flowed in the wind and obscured their faces.

"Yee ha!" cried the leader, leaping another timber breaker.

"Yippee!" cried his partner shrilly.

"Whoa!" then cried the leader, finding a path veering off the beach on to the field and slowing an enormous chestnut down to a trot, then together with his companion cantering slowly over to the small group observing them.

"Hallo there, you guys!"

"My God! It's Cecil! The young idiot! He could have broken his neck!" said Charlie, who always preferred anything but horses for transport.

Frank, Brab and Griffith grinned. It was typical of Cecil Grace, reckless as ever. But who was the young man with him? It certainly wasn't

Percy, his brother. Cecil leapt off his horse as Charlie backed away slightly with mild distaste.

"Why, good day to you folks!" Cecil held his reigns out to Frank.

"Would you mind holding my horse for a minute, Frank?"

He walked over to the black mare which had smoothly drawn to a halt. The mare shook her head and breathed steadily, hardly displaying any sign that she had just been raced for two miles along a sandy uneven beach with breakwaters.

Cecil stroked the mare's neck and whispered into her ear, then expertly took the reins and held out his hand so the rider could dismount.

It was only when the rider dismounted that the small group realized that Cecil's companion in wild horsemanship was a lady. Her long black hair had been tied back and she was wearing men's trousers.

"May I introduce you to Mrs Amy Cunningham!" cried Cecil almost proudly.

Frank bowed. "How do you do, ma'am! Frank McClean! Do you know, apart from my niece Sabrina, I have never seen a woman able to ride with such skill at such speed!" He smiled warmly and thought privately, "Well, well, Cecil!"

Griffith and Brab came forward and bowed in turn and introduced themselves.

"How do you do!" murmured Amy, and met their scrutiny without embarrassment. She had not ridden since she was a girl spending holidays with her grandparents on their farm in Yorkshire. Her grandfather had bred racehorses which he took regularly to York to run in the meetings there. All Amy's male cousins had learnt to ride on the horses in his stables and he saw no reason why his only granddaughter should not have the same opportunity. She was a gifted jockey and had an uncanny knack of handling animals.

"How splendid!" cried Brab. "What spirit, ma'am!"

Charlie came forward and bowed. "Good day to you, Mrs Cunningham! Charles Rolls!"

"Ah!" cried Cecil, retrieving the reins of his own horse from Frank, and expertly controlling both horses at once. "Mrs Cunningham lives at Shurland Hall, Eastchurch," Cecil informed Frank, Brab, Griffith and Charlie.

"Yes indeed!" replied Amy. Amy knew more about these men than they realized and she was aching to ask questions about their progress. Unconventional as she was, she could not quite bring herself to be so forward. "I'm so deeply impressed with your flying experiments." She hesitated. "Cecil has very kindly offered to show me your workshops. I do so hope it will not be a nuisance."

"No, certainly not," said Griffith.

"It would be a pleasure," said Frank graciously.

"Of course!" said Charlie rather stiffly.

Amy smiled radiantly. "Oh, how marvellous!"

All four men came forward to escort her to the sheds as if she were dressed perfectly normally in a white organza long frock and straw hat. They appeared to see nothing odd in this tiny woman dressed in men's riding clothes with an American hat hanging down her back. Cecil was left behind with the two horses, which he took over to tie up at the house.

"Do you keep a large stable, Mrs Cunningham?" enquired Frank, as they made their way to the first of the sheds.

"As a matter of fact, I do not," replied Amy. "Actually these horses belong to Mr Ingleton, our neighbour at Stone Pitts Farm. Cecil borrowed them. As it was such a beautiful day and the tide was out he thought it would be very pleasant to ride along the coast to Shellness, and indeed it was." She laughed, a light, tinkly laugh that enchanted her escorts.

Stopping at the entrance of the first shed, Amy turned to Frank McClean.

"You know, Cecil thought that it would be a frightfully good idea if the aeroplanes and gliders that came down before they should . . . so to speak," putting it as delicately as she could, "might be retrieved with the aid of horses rather than a dozen or so men. What do you think?"

"My word!" cried Brab. "What a splendid idea! Excellent! Using horses to rescue our machines is absolutely the answer."

Later, Eustace returned to London on the Sheppey Light and Oswald, after a modest supper at the public house called The Billet, opposite the Church, crossed over the road to return to his lodgings over the post office in Eastchurch, and settled down to write to Horace.

8th June, 1908

Dear Horace,

I hope I find you and the family well and enjoying this spell of fine weather.

I am sure you will have read the recent discouraging reports in The Times. Of course Eustace and I are firmly committed in our purpose, and the only thing that can possibly prevent us from forging ahead is the usual cash-flow business. We had a useful meeting today, mainly discussing the progress of our work schedules and some recent experiments which Eustace and I have been working on.

The subject of the War Office's claim that experiments with aeroplanes are unnecessary and impractical at the moment did not have any significant effect on the pioneers, apart from one of contempt! Nevertheless, I could not help my attention being drawn to the fact that they will not take any of us seriously until something spectacular occurs. I refer of course to a sustained flight or a flight from one side of the Channel to the other.

Eustace and I are hoping to visit France when Wilbur Wright returns there in the next few weeks to demonstrate his aeroplane. You can imagine how much we are looking forward to this visit. I will of course correspond immediately we return. Eustace and I and of course Mother are looking forward to your visit.

Your affectionate brother,
Oswald.

Oswald took a fresh page of writing paper and wrote,

PS A friend of Mr Cecil Grace has solved the problem of retrieving the gliders and aeroplanes from outlying fields. She has suggested that we use horses. She was rather an unconventional woman, she arrived on horseback after our meeting with Mr Grace, wearing men's clothing. I gather she lives at Shurland Hall, Eastchurch.

Oswald looked at the postscript he had written and decided on second thoughts to tear it up. Horace might jump to unfounded conclusions. Instead he wrote:

> We have resolved the problem of retrieving gliders and aeroplanes marooned in fields by using horses to pull them back.

CHAPTER TEN

Brab goes to France and Horace visits Leysdown

John Moore-Brabazon was the youngest of the pioneers. He was twenty-four in 1908 and a year younger than Oswald. He had married, in 1906, Hilda Krabbe, who was born in Argentina although she was English, of Danish extraction. She was as adventurous as Brab, and drove her own Panhard when they first met. At the time she was keen to acquire a Renault that would go faster than her Panhard. She was by all accounts the perfect partner for Brab, who was the most impulsive, high-spirited and optimistic of all the pioneers. It was with typical bravado that he swept off to France determined he would learn to fly, disgusted and outraged that the public and the press were so discouraging and unsupportive. On top of that he was challenged by a patronizing letter in *The Observer* remarking that "if pigs could only fly, then we would all happily follow suit".

As soon as Hilda and he arrived in France he made arrangements to buy a machine from the Voisin brothers, who believed that that he was a rich Englishman with no knowledge of mechanics. On both counts they were wrong. Brab had a good knowledge of mechanics, and when he reassembled his engine completely by himself, so that it ran better than it did originally, the Voisin brothers began to treat him with some respect. Brab's first flight took place in Paris in an early Voisin which he recalled

> was really no more than an enormous box-kite, with an elevator in front, an engine with a rather crude metal propeller, and a rudder in a colossal biplane tail. It had no side control, so if you

> tilted one side down you had to turn the rudder to force yourself back again on to a level keel. But as the turning of the rudder was in the opposite sense to the balance of a bicycle it was quite a tricky manoeuvre.

In due course Brab was very anxious to install a wooden propeller, and after much searching he finally found a factory which would take on this tricky task. They completed the job satisfactorily, but when Brab went to collect his propeller he discovered to his amusement that the factory's main occupation was manufacturing lavatory seats.

The interest and excitement among the inhabitants of Paris for early flying was like a tonic to Brab.

> I remember one day doing a flight of about one hundred and fifty yards at about fifty feet, which was unusually high because a gust of wind had hit me on the nose and sent me to my immense delight to this height. This was in fact the first time I was ever airborne in a heavier-than-air machine, and my excitement was genuine enough – as you may imagine – but not by any means free from fear. I describe the sensation afterwards as being like "sitting on a jelly in a strong draught" – a good enough description, but one that does little justice to the apprehension I felt. Loneliness, instability, uncertainty, the knowledge that no one in the whole world could help you – these were the sensations that combined to make one's heart beat faster.

When Brab landed, rather shakily, he was greeted by the crowd like a hero, and kissed warmly by shopkeepers, sales girls and taxi drivers alike, a very different response to that of the general public in England.

Subsequently they went to Chalons with Brab's *Bird of Passage*, where they met the famous Henri Farman. Brab practised flying constantly in Chalons, and often flew for as much as half a mile, but never more than about thirty or forty feet high. He brought the *Bird of Passage* back to England to show at the first Aero Show at Olympia in 1908. It turned out to be the only machine in the whole exhibition that left the ground. As usual only the *Daily Mail* seemed to show any real interest and it was here that Brab first came into contact with Charlie Hands from the newspaper, which offered a prize for the first all-English machine that could fly a mile. The seed of ambition was sown and Brab lost no time in approaching Eustace and Oswald to give them an order to build this machine.

Horace and his family duly arrived for their stay in London and Horace, wasting no time at all, made straight for his brothers' workshops in Battersea. The workshop was overflowing, and there was hardly room to move. Horace tactlessly remarked on the apparent chaos, which did nothing to improve his relationship with Eustace, and they immediately started arguing. Oswald had to step in to calm things down, and hastened to explain that they were in the middle of moving various works to Leysdown. Oswald offered to take Horace and the family down to the Isle of Sheppey and show him Mussel Manor and their new workshops in Leysdown that weekend. Eustace muttered that he was training for a boxing match.

In fact, Horace was deeply impressed, despite his criticism, but somehow could not bring himself to apologize for his inappropriate remark. He could see at a glance just how swiftly they had progressed from the early days, and when Oswald described to him the latest order that John Moore-Brabazon had given them and the immense challenge it presented, Horace was duly reduced to silence.

It was a hot summer's day when the Short family boarded the Sheppey Light train at Victoria Station, bound for Eastchurch. They were accompanied by day trippers from the East End, who were heading for Sheerness and Minster. Horace's children, Leonard, aged eight, Ruby, aged five, and the youngest, Francis aged three, were all excited at the prospect of a day by the seaside. Marie Claire had packed a picnic to eat on the beach at Leysdown. They climbed into one of the pony traps waiting outside Eastchurch station and instructed the driver to bring them to Mussel Manor and collect them again later on that afternoon.

Horace believed that the location was well chosen. He could see the advantage of the isolation, the golf course which could be used as a practice area, and the spot on which Oswald and Eustace were about to begin to build their sheds. Oswald had hired some local labourers, and had every intention of supervising and carrying out a great deal of the work himself. The accommodation at Mussel Manor would be useful and the fact that the village of Eastchurch was close, but not too close, was also an advantage.

Horace was thoughtful throughout the trip, digesting the huge scale of the possibility of heavier-than-air transport manufactured by his own brothers, who were clearly among the leaders in their field. It was obvious to Horace that they had no real concept how their ultimate ambitions would change the world for ever. He and Oswald walked along the shore towards Shellness and looked back over the fens, where there were no trees and no visible obstructions to speak of. On a clear day, there was an uninterrupted view of the Harty Marshes and the River Swale, forever changing colour in the reflections of the sky and clouds. The marsh birds called continuously to each other, diving in and out of the pools of water. Horace had already digested some of the extraordinary history of the Isle of Sheppey, and he was sure that, from Roman times, very little had changed the mysterious soul of the island. The legacies from the past were scattered around the villages, and left the island with ghosts of a mystic atmosphere difficult to fathom.

When the pony and trap came to collect the family they decided to have a look at Eastchurch and have a cup of tea before the next train was due. They passed Shurland Hall and soon came to All Saints Church opposite the Castle pub. The driver of the pony and trap told them that The Castle was where they could get a cup of tea. It was there that Charlie Rosewarne was having an early pint surrounded, as usual, by villagers enjoying one of his endless stories. Rosewarne immediately welcomed the newcomers and introduced himself. He could tell at a glance that these people were not ordinary trippers and he guessed correctly that they were something to do with the new tenants at Mussel Manor. Naturally the whole village already knew that Andrews, the builder from Minster, had had a lucky break. Rumours were rife, but the most popular one was that these rich posh blokes were going to build heavier-than-air machines. Neither Oswald nor Horace fitted into the category of rich or posh, but nevertheless they did not look like ordinary tourists. Horace soon sized Rosewarne up as someone who wielded some authority and was at the same time a popular resident of the village. He immediately introduced himself, much to the horror of

Oswald, who wanted to keep their identity secret. Horace told Rosewarne that he was from Newcastle and was visiting his brothers, who had just leased some land in Leysdown to build some workshops. Oswald was appalled at Horace's candour, and left the pub telling him not to say any more. Horace shrugged and told Rosewarne that he would soon find out what they were up to anyway. Rosewarne laughed and told Horace that they probably already knew because it had not taken long for the gossip to circulate. Tea was ordered for Marie Claire and the children, while Horace joined Rosewarne with a pint of the local beer. Horace explained that Marie Claire did not speak English, and that she was French, and from the Channel Islands. Rosewarne told him that his daughter went to a French convent in Sittingbourne and was always jabbering away in French to her mother for practice. The next time they came to Eastchurch they should come and visit them, and with that he gave Horace his card, saying he must be getting off home or he would get hell from his wife for being late for his tea. He downed his pint in one gulp, and left the pub. Horace wondered all the way back to Newcastle why Rosewarne thought there would be a next time.

CHAPTER ELEVEN

Charlie has lunch with Griffith Brewer

"So", Charlie questioned Griffith one morning in his office in Fulham, following the meeting they had at Mussel Manor, "what do you know about the people living in Eastchurch? Mrs Cunningham for instance?"

"Nothing really," replied Griffith, "except that she lives in Shurland Hall, once the home of the Earl of Shurland, the Cheynes, Henry VIII and Elizabeth I, and that apparently it's haunted and the villagers won't go near the place in case they see a headless Anne Boleyn."

"Where did you learn all that?" asked Charlie.

"The builder, who is also the coroner, a Grand Master Mason and head of the local skittles team! I believe his wife is the deputy headmistress of the local village school. Obviously respected people in the village."

"Hmm," mused Charlie, "interesting, my cousin is married to Lord Chief Justice Coleridge whose sister is Dulcie Dickson, the vicar's wife. Small world. The Dicksons suffered the most appalling tragedy about ten years ago: both their sons were drowned in a dreadful accident not far from Leysdown. I should go and call on them."

"You should," replied Griffith, "and you should try to meet Charlie Rosewarne. You'd like him; he has a good sense of humour."

"I shall call in to Eastchurch next time I visit Leysdown. Now, Griffith," said Charlie, picking up a society magazine and opening it at a page illustrating a Silver Ghost motor car, "what do you think of this?"

> Those who try the Rolls-Royce buy the Rolls-Royce
> Those who buy the Rolls-Royce will buy none other than
> the Rolls-Royce
> And give repeat orders for the Rolls-Royce.

This was the advertisement that Charlie Rolls published in society magazines and *The Motor*.

"Not the greatest literary piece, but if it does the trick, then what does it matter?" said Griffith.

Despite its lack of literary brilliance, it was effective, according to Claude Johnson, Charlie's business partner, who saw a definite conflict of sales strategy between racing cars and touring cars which might very well have hampered the popularity of Rolls-Royce cars and wrote in the *Car Illustrated*:

> It is my humble opinion at the present time that the Automobile Club should do everything in its power to prevent the high speeds of automobiles being brought before the public notice, whereas at the present time it seems to be devoting too large a part of its energies to exhibiting not only racing cars but touring cars driving at their maximum speed.

Although Charlie was no longer racing himself, his interest in racing and rallying was still very strong, and the Rolls-Royce success in these fields was very important. He was determined to succeed in making his Rolls-Royce company a worldwide name. He travelled for miles testing his cars and demonstrating them to clients. Royce was grudging in his appreciation of the work that Charlie did on their behalf, often complaining that he did the real work. But of course this was not true. It was Charlie's enormous practical experience as a racing and rally driver that enabled him to pick up the design and engine faults which Royce would never have found. There was no doubt that Charlie was responsible for many of the design features for which he was never given proper credit. For instance, Charlie introduced the adoption of wire wheels as a much more reliable design than the existing structures. Royce objected to the cost, claiming that he did not think that the expense of such a design was warranted.

Johnson, respecting Charlie's experience and astuteness, offered to finance the improvement from C.S. Rolls and Company. Royce gave in, and wire wheels were adopted from then on. But it was Charlie's practical approach that ultimately made the taciturn Royce grudgingly recognize Charlie's long experience and understanding of the motor car. Although they were opposites in many ways, the one thing that united them was their passion for perfection in the manufacture of their Rolls-Royce cars.

Charlie never sought recognition of his ideas or input, as he was so committed to achieving perfection. He was one of the greatest entrepreneurs of his time, but his complex personality was always at odds with convention, and his extraordinary foresight never received the recognition to which he was entitled. Brab described his outlook as "almost prophetic". Unlike prophets, however, his vision was all based on practical experience, and by 1907/8 the revolution of the motor car had passed into evolution, and the so called daredevils of the road were turning their attention to the air.

Charlie's focus was shifting to flight, despite the demands made of him by his business. His visits to Manchester, his London showrooms and workshops were never properly scheduled, so that he would appear unannounced, disconcertingly, like a whirlwind, taking on board exactly the progress, or lack of it. He would evaluate what needed to be attended to, give concise instructions and leave abruptly. He often forgot to eat regularly and seemed to exist on very little sleep, but was sharply aware of opportunities which might benefit the company.

Airships had never interested him but nevertheless, when he saw an opportunity to build the casing for one he quickly secured the contract to build it, and rather against his intuition that airships were impractical, he still needed the experience of an ascent to satisfy his flying curiosity. He managed to fit in a trip with Frank Hedges Butler and Mrs Assheton Harbord in the French airship *Ville de Paris*. Charlie was excited enough to compare the experience with a Jules Verne or H.G. Wells adventure, but it confirmed his opinion that airships were not a satisfactory option, and he voiced his reservations in the *London Magazine*:

> . . . if a propeller-shaft broke – and when one contemplates the flimsy assortment of shafting that adorned early dirigibles, one recognizes

> this as a likely contingency – the propeller would be carried away. Not only would this deprive the airship of control, it would also deprive it of precious weight, and the craft would rise sharply, while the gas expanded and leaked out through the valves. As it re-entered the clouds, the gas would contract, the envelope would grow heavy with moisture, and the structure would fall with terrible swiftness. The weight of the structure was too great to allow the crew to check the descent with a normal load of ballast. The frame of the car would break on impact with the ground, and the lightened ship would rise again, only to descend still more violently. A minor engine disorder would reduce the airship to the uneasy status of a balloon, and the control car was too light to protect the occupants from the consequences of a really hard landing. The only hope for the aeronauts in such a hypothetical case would be to land on a wood.

However, Charlie was undeterred from taking on the manufacture of the drive shaft and casing for Colonel Capper's Army airship at Lillie Hall. The story is told by Harry Fleck, one of his mechanics,

> Charlie got all the blueprints and dimensions, and shafting and drives were made and the aluminum casing. This was a T-shaped tube suitably flanged with bolts etc. It had to be inspected by C.S.R. It was laid out in the centre of the hall and bolted up so that Charlie could make his inspection. Everything was O.K., and Charlie duly informed. Some time afterwards one of the lads got a piece of brown paper, pushed it into the tube, and lit it with a match. It put up some good smoke, and looked very much like a ship stoking up, and all hands were laughing their heads off when Charlie drove in and saw the whole thing. He flew off the handle on the Saturday, and came in on Monday as though nothing had happened, and we never heard another word about it.

If Harry Fleck and his friends were in trouble, it was nothing compared to the fury of the Board of Directors that was levelled at Charlie for taking on a contract without their permission. Neither Royce nor Johnson ever supported Charlie in his aeronautical experiments. This was a particularly pertinent example of Royce's limited horizons. From then on Charlie never attempted to promote his aeronautical ambitions into Rolls-Royce Limited, but went full steam ahead without involving the company in any way.

"Come on Griffith, let's go for a spot of lunch!" said Charlie.

"Good idea," replied Griffith, surprised. Charlie rarely volunteered to go out for lunch "Where?"

"Eustace Miles; it's a splendid vegetarian restaurant," exclaimed Charlie.

Griffith was well aware of Charlie's food fads. Charlie was the treasurer of the National Food Reform Association, whose aims were to reform diet in institutions "and recommend more hygienic, humane and scientific methods of selection and preparation of food".

Charlie kept a strict "keep fit" routine to accommodate his overburdened schedule which was not entirely governed by regular exercise, but rather by a strict diet. He believed seriously that the general public should reform their food habits. He was an avid supporter of the National Anti-Vivisection Society, although he was by no means a vegetarian, and although he rarely drank he was by no means teetotal. In an article in *Pearson's Weekly* in 1908 he wrote:

> I have never had flu. The method is to stop all unnecessary internal work and waste by cutting off all the food that is not actually required for the proper repair and sustenance of the body; all surplus food beyond this means the infliction of unnecessary labour on the digestive and other organs at the expense of the system. By confining oneself to two light meals a day – well masticated – the human machine, being no longer handicapped in this manner, is thus able to confine its whole energy to driving out the disease.

Griffith was at pains not to mention the subject of Vera, who was now married. The rules of society in 1908 were cruel and although Griffith was sure that Lord Llangattock would not have stood in their way he knew that Lady Llangattock would have been a different matter, she would not have welcomed Vera as a daughter-in-law and she would have made sure that Vera would have been ostracized by society. Lady

Llangattock had an iron will and a resistance to showing any emotion whatever the circumstances, and considered that doing so was vulgar in some way. There was virtually no hint of any sort of lasting relationships with the opposite sex as far as Charlie's siblings were concerned. Charlie had rooms in his parents' house, South Lodges, in Kensington but rarely encountered the rest of the family. His brothers and sister also had their own rooms there, and led completely separate lives, rarely coming into contact with each other.

Lord Llangattock was devoted to Charlie, and was immensely proud of his eccentric son, and Charlie appreciated his father's love, valued his advice, and probably confided in him more than in any of his friends. He understood that his father, unlike his mother, was not a slave to convention and that he did not care that Charlie might be considered to be "in trade"; nor did he care that Charlie had little in common with most of his contemporaries. At the same time, he was immensely impressed with the friends to whom Charlie had introduced him. But then, who could not warm to men like John Moore-Brabazon, Frank McClean and Frank Hedges Butler, who were the most entertaining and amusing company and a very good influence on his son.

Charlie's years of solitude as a child spent with the gardeners, nanny and tutors at South Lodges had given him a reasonably happy outlook on life, despite the absence of his parents for months on end. The Hendre, far away in Wales, was his home where the staff had loved him, indulged him and encouraged him. Charlie appeared to have inherited his mother's iron will but none of her restrictive social behaviour.

Charlie was not short of female admirers, and the Honourable Mrs Assheton Harbord was one who was immensely rich, socially acceptable, intrepid, a fellow aeronaut and a very good customer. Charlie was flattered by her attention, not to mention delighted with her influential connections, which supplied him with eager customers. It must have been heartbreaking for poor Vera, who had to watch Charlie's exploits from a distance, when once she had played a part in them. The beautiful Vera, amusing and adventurous and possessing all the talents which Charlie admired, could no longer fulfil the same role as she had prior to her marriage with Iltid Nicholl. History does not relate exactly how Charlie felt about her marriage but one suspects that, whatever he felt, he would not have betrayed his feelings either to Vera or to anyone else. But history does relate that Charlie was the unrequited love of Vera's life. In contrast, the Honourable Mrs Assheton Harbord, like a battleship in full sail, took Charlie under her wing, smartened him up and introduced home comforts into his apartment in South Lodges, which she observed was more suited to a hermit. She became a regular and intimate companion. She was obviously an extremely forceful personality but on the other hand had a generous heart. She meant well!

Griffith decided over coffee that he would take a gamble and threw the information that *Aida* was being performed at the Royal Albert Hall that week; perhaps Charlie was going to go?

"Mrs Assheton Harbord has certainly purchased tickets, but frankly, Griffith, I cannot abide opera and I have every intention of making my excuses and slipping off to The Players' Theatre where Marie Lloyd is performing."

Griffith laughed. "You cannot blame her for trying, Charlie!"

Charlie gave a wry smile. Actually, he was, himself, very clever, as everyone knew, including Griffith, at adapting imitations of music hall favourites to serve all purposes of criticism of car developments; for instance, his idea of conveying complete satisfaction of a new brake scheme. Pulling up the car in front of Mr Royce, in answer to the anxious engineer's enquiry for his opinion, Charlie would begin a recitation in the voice and actions of the late Dan Leno, starting with the words "Call that a brake?" Poor Henry Royce could never get used to Charlie's leg-pulling. Charlie's complex and sensitive character always surprised his many jealous critics, who were very often the subject of his rather wicked mimicry.

Griffith and Charlie finished their coffee. Griffith knew he would have to settle the bill. He didn't mind, Charlie's reputation for meanness was legendary, but he was always generous to children and Griffith knew that Charlie had often slipped his own children the odd shilling, which they had to earn by reciting something or winning a game or competition. Children loved him; he never talked down to them and his adventurous exploits in motor car racing, ballooning and aeronautics thrilled them.

CHAPTER TWELVE

Wilbur Wright returns to France

In September 1908 Wilbur Wright returned to France while Orville remained in America to carry out demonstrations there. Crowds flocked to Le Mans, astonished at Wilbur's flights, which were achieving an ever increasing time and distance. The more successful he became, the more crowds and onlookers increased, while Le Mans became more restricting, and so he made the decision to move to Auvours.

Charlie Rolls, Frank Hedges Butler and Griffith Brewer were already in France at this point for a Roads Congress, but as soon as they were able to leave that event they made their way to Auvours to watch Wilbur's demonstrations.

Now confident of his aeroplane and flying skills, Wilbur began to take passengers. Charlie was among the first to volunteer. He summarized his experiences in *The Times* as follows:

> Once clear of the ground, the feeling of security was perfect, and I was able to watch with great interest the movement of the operating levers. We tore along at a speed of 40 m.p.h., and soon came to the first corner. Here a point of interest to motorists was demonstrated, namely that, no matter at what speeds a curve is taken, the machine "adjusts its own banking", so to speak. At the will of the operator, it tilted up gracefully when taking a turn, demonstrating that taking a curve on a flying machine will, instead of being more dangerous, be actually safer than in an automobile. On this occasion our flight was more than usually interesting by reason of some strong side gusts that attacked the flier on certain

> parts of the course. The prevalence of these and other undulations of the atmosphere rendered very close attention necessary to the two levers controlling the equilibrium and altitude: I noticed that both these were kept constantly on the "joggle" with slight movement. One has been accustomed to consider the atmosphere as a mass of air, increasing in density with its altitude, but otherwise uniform. Experience on a power-driven flier, however, shows that, far from being the case, the atmosphere near the earth's surface – even in what we call calm weather – is made up of spiral movements of varying diameter, sometimes vertical and sometimes horizontal, undulations of all sorts, little hills and valleys, and 'streams' of air: in fact one might call it a new world conquered by man – a world with scenery of great variation which, though invisible to the eye, is none the less felt by the operator of a flying machine. To maintain equilibrium and steering control while battling with these complex movements of the air has been the great problem which for centuries has baffled human integrity and which is now solved by the Wright brothers after years of systematic study and experiment.

Charlie was tremendously impressed with the skill with which Wilbur manipulated his controls and studied his actions carefully:

> Sometimes we flew above the trees, sometimes we flew three feet off the ground, entirely at the will of the operator, who thus showed the most perfect control over his machine that anyone could imagine.

Charlie returned to England with Frank Hedges Butler, and although he could have chosen to enrol in Wilbur's flying school, he chose to teach himself on a glider he had commissioned Oswald and Eustace to build. He needed to find a suitable spot away from the public gaze to experiment. Charlie was always careful and circumspect, even in his early motoring days; he never rushed into anything of which he was not absolutely sure. So he took his time not only in choosing the design of the glider, but also in his reconnaissance of where he might go to teach himself to fly.

Griffith Brewer stayed on in Auvours for a few days to talk privately with Wilbur. Wilbur and Orville had not found a suitable manufacturer who they felt was reliable enough to enter into a contract. So far they had been unable to trust anyone. It was Griffith Brewer's golden opportunity to recommend the Short brothers and he suggested that Oswald and Eustace come to France to see Wilbur's demonstration. Wilbur trusted Griffith Brewer's judgement and, possibly due to the fact that there was no language misunderstanding, he agreed that a meeting would be beneficial to everyone concerned.

Griffith Brewer's account was recorded in *Flight*, years later, on 20th April, 1939.

> When Wilbur Wright came to France in 1908 and demonstrated to a disbelieving world that flight was possible, there were, naturally, no aircraft factories. There were, however, some aeronautical constructors who supplied the very limited demand of a handful of balloonists who required balloons to race amongst themselves and to take countries overseas to complete for cups offered by enterprising sportsmen.
>
> There was no thought of gain amongst the balloonists in the beginning of the present century.
>
> It is to the Hon. Chas Rolls that the credit is due for "discovering" Short Brothers. Eustace (then in his early twenties) and his brother Oswald, still in his teens, composed the aeronautical firm of Short Brothers, whose chief asset and guarantee for their future was a genius for excellence in workmanship, and an industry and perseverance which would carry them through enormous difficulties.
>
> My first personal contact with the Short Brothers was when they reconditioned the balloon used by Santos Dumont in the Gordon Bennett race in 1906, ready for me to pilot with my aide, the Hon Claude Brabazon, in competing for the Gordon Bennett Cup from St Louis in America in the following September, 1907. Later, when Shorts built the Bee for me, an 18,000 cubic feet rubber fabric hydrogen balloon, I was better able to appreciate the meticulous care and skill in their manufacture. It was about this time the Short Brothers were awarded the first prize at the Aero Club Exhibition "for excellence of construction".

> These were my early experiences with the Shorts, and it therefore was not surprising that when I became associated with the Wright brothers at Pau at the time when Wilbur Wright was making his flying demonstrations, I should suggest their name to Wilbur Wright when he enquired which firm in England would be most suitable to manufacture the first six flying machines on behalf of Wilbur and Orville Wright. I naturally replied that the only firm with any experience in aeronautics at that time, namely in January 1909, was Short Brothers, and at Wilbur Wright's request I telegraphed to them to come over to Pau to take particulars for making the first six flying machines to be used in England. This was the first aircraft order given up to that time, and in undertaking it Short Brothers became the first firm of aircraft manufacturers in Great Britain.

Griffith returned to England a few days later and went straight to Battersea to see Oswald and Eustace to tell them that Wilbur and Oswald Wright were interested in meeting them as soon as possible, and asked them if they would be available to come to Auvours by the end of the month. With the excitement and anticipation of witnessing Wilbur's flying demonstrations, and the opportunity to meet the Wright brothers, they readily agreed, and were grateful to Griffith for arranging a meeting that both Eustace and Oswald knew could lead to extraordinary results. Sadly, there is no proper record of this historic meeting but there is no doubt that the Short brothers were completely clear in their own minds for this next step. The detrimental reports in the newspapers and the challenge thrown down by the Admiralty Department that they and the Aero Club would not be taken seriously until "a flight crosses the Channel" made them all the more determined to manufacture an aircraft that would be capable of such a feat. They knew that what they were going to witness was going to change the course of history for ever, and they were absolutely determined on being part of this evolution.

History does record that the meeting was a success, and as soon as the two sets of brothers met, the Shorts' diffidence disappeared and their confidence and ability in their own skills was evident to the Wright brothers. They soon discovered that although their backgrounds were entirely different, both their early inventive beginnings were very similar.

During their long discussions Wilbur and Orville inevitably learnt of the hardships that the Short family had endured in their early life, and were deeply impressed with how they had overcome the obstacles through their own tenacity and vision. The fact that people like the famous Honourable Charles Rolls, John Moore-Brabazon and of course their good friend and agent Griffith Brewer had entrusted them to become the official Aero Club manufacturers, despite their youth, was impressive. The Wrights were cautious, they were still not ready to make a decision on a manufacturer for their own aeroplanes, yet they felt confident that here were the first trustworthy engineers with the ambition and ability to be able to produce an aircraft which would best meet the criteria. The two sets of brothers began a friendship which endured for the rest of their lives.

As soon as Eustace and Oswald arrived back in England, Oswald sent a telegram to Horace to say that he was catching a train to Newcastle that day to see him, and would be obliged if he would meet the train. He knew that, despite Eustace's reservations, they desperately needed their brother Horace. He was the only engineer Oswald knew who was not only well equipped, but also one he could trust implicitly to translate rough sketches into clear technical drawings, which he and Eustace could follow precisely in the manufacturing of aeroplanes.

Horace was naturally delighted that "The Kid" was making a flying visit but was quite unprepared for his burning motive. He was, of course, quick to assess the urgency of action to secure an unassailable place in aeronautical manufacturing. In fact, Horace had done little else, since his visit to Leysdown, but think of the extraordinary future of aeronautics that lay ahead. He knew immediately he must join his brothers. There was one problem: he was still under contract to Sir Charles Parsons; but he told Oswald he would go and see Sir Charles the following morning. Oswald returned to London the following day to wait for Horace's ultimate decision.

Sir Charles Parsons was well aware of the success of Eustace and Oswald's aeronautical business, and was full of admiration that the Aero Club had appointed them as its official engineers. It so happened that he was a friend of Lord Northcliffe, and shared Northcliffe's view that the Government and the public were wrong in their assessment of the flying pioneers, and should support them rather than denigrate them. He realized it could only be a matter of time before Horace would join them. Naturally, he hated to let Horace go but he was the last person to stand in the way of progress, and their need of Horace was indeed greater than his at this time. He had already benefitted hugely from Horace's contribution to his company, through the acquisition of Horace's Auxetophone patent, and the development and ultimate sale of the turbine engine to the Navy through a typical prank of Horace's so that, of course, he could not stand in Horace's way. He not only wished him luck but offered his future support and help if Horace needed it.

Horace telegraphed Oswald immediately: "I will give you and Eustace three days to make your minds up to proceed with aeroplane manufacture, and if you haven't done so by then, I will start without you."

A typical Horace pronouncement, which irritated Eustace, but he grudgingly accepted and acknowledged that Horace, despite the fact that he was their brother and someone who gave him more grief than anyone he knew, was nevertheless an exceptional engineer, and he knew that there was no other engineer with the capability or talent to produce the accurate technical drawings required. Nor indeed would he trust anyone else, if the truth were known.

True to his word, Horace broke the news to Marie Claire, Leonard, Ruby and Francis that he was going to set up a company with their uncles, and Marie Claire's brothers-in-law. Naturally Marie Claire was shocked while the children did not know how to respond. Horace told them he had to leave immediately but he would be back as soon as he could make arrangements for them to join him. Francis's only concern was that Gracie, their bull terrier, could come with them. Francis was five years old and like his Uncle Oswald adored animals.

"Naturally," said Horace laughing. "She's part of the family."

Francis, at least, was happy and excited at the prospect of seeing more of his uncles. Leonard and Ruby were dismayed at losing their friends and Marie Claire hated the idea of the disruption of going to a new place where no one would understand her. At least in Northumberland she had found some fellow French people.

Oswald, despite the fact that he was still only twenty-five years of age, quietly took control. He was determined to make their partnership strong and enduring so that their separate talents could be exploited to the full. When Horace arrived in London they arranged to meet at their mother's flat at Battersea Park. To be fair, the three brothers were determined that this opportunity to work together and use each other's talents could change the course of aeronautical history for ever. They were determined, whatever their personal disagreements might be between themselves, that they would always be loyal to each other and respect the integrity of the company they had formed. They soon came to an agreement together and set up their company as "Short Brothers" with a capital of £600, each contributing £200.

The Aero Club were absolutely delighted with the news that not only had Eustace and Oswald's brother Horace left Northumberland to join his brothers but that they had set up a company together and furthermore they had embarked on the manufacture of their first aeroplane. Horace was using his mother's drawing room as his drawing office. Griffith Brewer was certain that now Horace had joined Eustace and Oswald the Wrights' potential contract of manufacturing six Wright Flyers was almost assured. But he knew that the Wrights were cautious by nature and would not embark on a contract unless they were absolutely sure that their decision was going to work. Griffith hoped that when he told them that Horace had now joined his brothers they would feel confident to proceed.

Oswald, Eustace and Horace's first customer was Frank McClean, who placed his order aboard a ship bound for China to witness the eclipse of the sun. The cable merely said "build me an aeroplane".

No price was specified. Oswald wrote later: "Horace designed every detail of the machine himself, with the drawing room of my mother's flat at Battersea Park as a drawing office. The machine was partly built in our balloon works, and it was exhibited at the first Olympia Aero Show in 1909 in an unfinished state. It was finally erected at Leysdown."

The aeroplane was called Short No. 1 but sadly it never flew as the 40hp Nordenfelt engine taken from a motor car was much too heavy. But the Shorts learnt a great deal from this, their first excursion into the manufacture of an aeroplane, so that Short No. 2's manufacture was started shortly afterwards, commissioned by Moore-Brabazon.

Horace wanted to move his family south before Christmas despite

Marie Claire's objections that the children would miss out on the festivities with their friends over the Christmas period. Horace was determined to get them settled as soon as he could. Marie Claire knew it was pointless arguing with Horace. His temper was something she could not cope with, and they were all afraid of him when he lost control.

Horace felt that Eastchurch would be a good place to live. Close to Leysdown and a station from which you could catch a train to Victoria. He went in search of Charlie Rosewarne, the builder, whom he had met on his previous visit. He felt sure that Rosewarne would know of suitable properties available for rent. He drew up in the middle of the village and found the post office. He caused quite a stir: motor cars were rarely seen in the village although recently several had passed through on their way to Leysdown. Customers in the post office were duly silenced as Horace asked if they knew where he might find Charles Rosewarne. The postmistress directed him to Shurland Hall. He found the property quite easily and was quite surprised that such an impressive property should exist in such a small village. It was clearly a very closely knit village, and the lady in the post office was probably aware where everyone was at any given time. Horace was sure that news of his arrival in the village inquiring after Charlie Rosewarne would soon be passed on.

Horace walked into the courtyard of Shurland, which he realized dated back to pre-Tudor times. He rang, or rather clattered a bell on a massive oak door which was opened by the diminutive Mrs Cunningham.

'Ah," exclaimed Horace with the twinkle in his eye whenever he engaged with a pretty woman. "My name is Horace Short. I was looking for a Mr Charles Rosewarne who I was told might be working here."

"You're in luck," twinkled Amy in return. "My name is Amy Cunningham. How do you do? Do come in!"

Horace stepped into a wonderful reception hall with a massive oak staircase running up to the first floor, which separated halfway up and arrived on opposite sides of an impressive landing.

Rosewarne emerged from behind the staircase.

"Mr Short!" he proclaimed. "I had no idea I would see you return so soon. I have a few more things to discuss with Mrs Cunningham. Would you like to meet me in half an hour at The Castle?"

"Perfect!" declared Horace. "Good day to you, ma'am," he said, bowing to Amy Cunningham. "I hope I shall have the pleasure again soon."

"I hope so too," replied Amy politely and ushered Horace out of Shurland Hall.

This gave Horace an opportunity to look round the village before his meeting with Charlie Rosewarne. On his way back to the village he passed a pretty duck pond and some picturesque cottages, the village school and the pub, The Castle, which was almost opposite the other village pub, The Billet. The Castle was separated from the church by a narrow lane which led to Leysdown, approximately three miles away. All Saints Church occupied the centre of the village and was, according to the numerous notices in the porch, the central focus of activities in the village. From the information on the notices Horace learnt that the Reverend Dickson was the incumbent vicar, and his wife Dulcie ran various societies including the Flower Club and the Mothers' Union. Mrs Rosewarne also appeared to be heavily involved with village activities; she ran the Sunday School for the village children and was, unusually for a woman, head of the Parish Council. Clearly a force to be reckoned with. Horace recalled that she also taught in the local village school and, interestingly, Rosewarne's name appeared on a notice which suggested that he was the local coroner. Mrs Cunningham ran the choir and he noted that there were at least two choir practices per week and at least twenty people in the choir, of whom Rosewarne was one. It seemed that Shurland Hall was host to local events such as fairs, gymkhanas and agricultural shows, which all took place on Shurland Meadow. It was a typical English village.

The church itself was originally Saxon but had been rebuilt in 1431 by William Cheyne under licence from King Henry IV, who gave three roods of land for the site of the present church. Chalk boulders were imported by barge to Harty to give the church solid foundations, and a Cistercian monk, William Nudds, of Boxy, supervised the building and afterwards became the first vicar. The sole relic of this ancient church is an old alms-box, a heavy iron-bound construction with three separate locks. Apparently the vicar and two churchwardens each keep a key. The oak pulpit was built in the reign of James I. Among the spectacular windows is one which commemorates one of the two sons of the Reverend Dickson who were tragically drowned in 1890. The most attractive tomb in the church is that of Sir Gabriel Livesey of Parsonage Farm and his wife Anne. Above the canopy of the tomb is a grinning black head, in remembrance of Sir Gabriel's black servant. The small kneeling figure in the front panel is believed to be his son, the notorious Michael Livesey, whose name appears on the death warrant of Charles I. Sir Michael Livesey by all accounts had a very unpleasant reputation, as having accepted a knighthood from King Charles I he sided with the Roundheads and sat in the court which authorized King Charles's execution and added his name to the death warrant. His behaviour as a soldier was apparently far from brave and stalwart; he was a bully and a proven coward on the field of battle, but his huge wealth and power seemed to redeem him in some bizarre way.

There was a forge on the other side of the church, and the vicarage was just behind that. Opposite the church was a lane, Church Road, which led down a hill signposted "to the station". Walking a little way down this road, there was a fine view of undulating pastures and fenland with the River Swale darkly shining in the distance. Horace caught the mystique of the island, and contrary to advice that he would be better off living in Sittingbourne on the mainland, it confirmed his view that this was where he wanted to be.

Later, in The Castle, he explained to Rosewarne that he was joining his brothers and that they had formed a company together, "Short Brothers". He told him he liked what he had seen of the island and he particularly liked Eastchurch, which was conveniently close to Leysdown, and he was sure that his family would be happy here. Rosewarne was delighted to hear the news; it was good for the village to have new people in the community. He told Horace that he knew just the place. The owner, Mr Taylor, was a friend of his and he was sure Mr Taylor wouldn't mind if he showed Horace the property, and if he liked it Horace could call on him to negotiate a lease. The property, Parsonage Farm, was a mile or so down Church Road, and was the birthplace of the notorious Sir Michael Livesey.

Parsonage Farm was exactly the sort of property that Horace had in mind, something with plenty of character, plenty of outbuildings and plenty of space. There were stunning views over Standford Hill and the Swale, but an easy walk to the village. Horace signed a lease which allowed him and his family to occupy the property in December 1908.

In some ways Horace was sorry to leave Northumberland, because he knew it would be the end of a quiet, stable family life, probably for the rest of his days. Not that he looked forward to the move with any regret whatsoever, as he could barely wait for the challenge he faced. Sir Charles Parsons had wished him luck and warned him that the biggest battle he would face would not be his own ability to manufacture the best aeroplanes, but to persuade the Government to take them seriously. Horace remembered all too well how the obstacles made by so-called intelligent men had restricted his own progress in the past. Sir Charles Parsons' advice was to find a sympathetic, enthusiastic and influential ear. Unfortunately at that time he could not recommend anyone.

CHAPTER THIRTEEN

Horace Short's family move to Eastchurch

By the time Horace arrived at the outskirts of Eastchurch with his family, it was getting on for half past three in the afternoon and the December light was already beginning to fade. They had left Northumberland very early the previous day and spent the night at his mother's flat. In the morning Horace had taken the children to London Zoo as a sort of treat for missing out on a Christmas with their friends in Northumberland. Leonard was ecstatic and vowed that he would become an explorer in Africa and live with the chimpanzees there. Francis wanted to free all the animals from their cages, except possibly the snakes. Marie Claire reorganized the packing and spent the morning with her mother-in-law. She had softened slightly and resigned herself to yet another change and move in her life.

The last red rays of winter sun were sinking over the Swale as they drove into the yard of Parsonage Farm. Gas lamps had been lit in the front room. Horace turned the engine off and opened his motor car door. Gracie leapt over him and out of the car, sniffed the air and barked. The front door opened and an identical bull terrier came bounding past Charlie Rosewarne, standing on the doorstep to greet them. His wife Kate had had the house aired and cleaned and had stocked the larder with essentials. Rosewarne insisted that they all come to Rosebank for supper and to meet his wife and daughter, Dorothy.

"It's only a short walk up the hill," said Charlie. "Bring your dog; looks like they're friends already!"

The Short family followed Charlie Rosewarne a hundred yards

or so up Church Road and turned right into a building yard with several workshops and a timber store. They walked through the yard and through a small gate into the back garden of Rosebank and entered the house through a small glass porch into their dining room. There was hardly room to move; the room was dominated by a huge ebony dining table, sideboard and upright piano. A fire was burning in the grate and oil lamps gave a cosy glow.

"Sit down, sit down everyone, tuck in," bellowed Charlie Rosewarne.

Dorothy, eight years old, the same age as Leonard, could hardly believe her eyes when she saw an identical bull terrier arrive with the family. What made the evening for Marie Claire was the fact that Kate and Dorothy both spoke French. Dorothy liked Leonard, Ruby and Francis, especially Francis, even though he was the youngest; she recognized a kindred spirit immediately. She promised to come and see them the next day and show them around the village.

"I'll bring Henry with me," she called. Henry was the name of her bull terrier.

It was still dark when Horace left Parsonage Farm on his first morning in Eastchurch, and there had been a thick frost. His footsteps across the front lawn left their imprints, and his boots scrunched loudly on the gravel as he made his way towards the barn where he had parked his motor car. He stopped and turned to look across the fields in the direction of Standford Hill. He could just make out the Swale in the distance and the early streaks of dawn glistening on the coal black river water, as the winter moon faded at the prospect of the impending day ahead. The white frost spread across the fields like icing sugar, and it seemed to Horace that these could make perfect flying fields. He jumped as a marsh bird awoke and screeched sharply, breaking the peaceful silence of the dawn.

Horace turned around once again and walked into the barn. He fired up the engine with one turn of the starting handle, drove out of the yard and turned left on to Church Road, up the hill towards the church, where he turned right towards Shellness and Leysdown. He drove past Shurland Hall, looming darkly and forbiddingly amongst the white frosted fields.

How much land did the Cunninghams own? He had heard of the collapse of John Cunningham's merchant bank and he had heard about his drinking problem. Little wonder that Amy enjoyed the attentions of the glamorous Cecil Grace. But he did not dwell on her for long, for his mind was racing with the possibilities and challenges which his new partnership with his brothers had brought.

He had met Frank McClean before, but only briefly. It seemed to Horace that his order for an aeroplane was the most magnificent gesture of faith they could have hoped for. He had placed the order for the aeroplane by letter, while aboard ship en route to China to witness an eclipse of the sun. The gist of it was simply "build me an aeroplane", with no tender price sought.

Oswald was waiting at Mussel Manor when Horace arrived. He had stayed there overnight to save time. It was a spooky place to stay, and extremely uncomfortable. Oswald was as excited as Horace at their new partnership.

"Well, Kid!" said Horace with a broad grin; "let's get going!"

It was now seven thirty in the morning and men were arriving for work. Some had walked for miles from Minster and Queenborough and others had come by bike. The gas lamps cast long beams of uneven light in the sheds, while a coal stove spluttered to life in the corner. There was the usual banter amongst the craftsmen as they resumed their work from the previous day. One man announced he had become a father in the middle of the night; another that his wife had locked him out after staying too long in the pub. Cheers and gales of laughter resounded around the shed in response.

When Horace and his family had moved into Parsonage Farm in mid-December 1908, the workshops at Leysdown were under way and being closely supervised by Oswald. They consisted of four constructions made mainly of steel with high, sloping, corrugated iron roofs strong enough to withstand the winter gales. They were going to be twice as big as the ones at Battersea and able to hold a work force of up to fifteen men each. Oswald was delighted that Horace had chosen to live in Eastchurch, only a few miles away. Although it was practical for Oswald to stay at Mussel Manor on occasions to be on hand for the organization of the building of the workshops, he found it very lonely and extremely eerie there on his own. The dark, low-beamed ceilings and flagstone floors all added to the ghostly atmosphere. It was freezing cold, there was no electricity and the candles Oswald used to light his way to bed cast mysterious shadows, giving his imagination some nasty frights. There was not another house in sight, and sometimes on dark and foggy nights it was impossible to find one's way, as there was no real road to the works. One such night one of the workmen did in fact get lost on the marshes and could not find his way back. In Oswald's own words:

> Mussel Manor was a small ancient house about half a mile due north of our factory. It was taken over by the Aero Club as a clubhouse but it was seldom used except for an occasional lunch by a few members who were interested in flying. It had only two medium sized rooms which could be used by members of the Club, and a few bedrooms. Mussel Manor was occupied when we arrived by a caretaker, an elderly woman who lived alone. It was she who prepared lunch for several members of the club when it was necessary but the clubhouse was little used for this purpose . . .
>
> During the earliest days at Leysdown, I was only there occasionally, as Eustace and I were still much occupied with our work at our Battersea balloon works and there were still a considerable number of balloon ascents being made by members of the Aero Club.

The order books were full and the extra space away from the public gaze was ideal, but unfortunately Oswald found that he had become

> the chief and only book-keeper, an irksome task because I was not inclined that way, and consequently found no pleasure in the work. Apart from that, the proceedings at Leysdown were full of interest and occasionally exciting. There was still work to be done with Eustace at our balloon factory, and an occasional balloon ascent with myself in charge of the balloon as pilot.

Oswald and Eustace had already experimented with building a glider for Moore-Brabazon and Cecil Grace but, as Oswald said:

> this glider never glided . . . I took the bits and pieces of that glider to Lewes, where Moore-Brabazon and Cecil Grace were staying, and I erected it in one day in the Castle grounds on a boiling hot day. I attached a long length of "bungee" to give the machine an assisted take-off. There I left it in Moore-Brabazon's care . . . I had quite forgotten the existence of this would be glider, which had such an affinity with Mother Earth, and it took me a long time to find out how it became dignified with the title of "airplane". It appears that Moore-Brabazon with the assistance of Warwick and Howard Wright installed an engine and propeller in it, hence it became an aeroplane.

But Oswald and Eustace built a much more successful glider for Charlie Rolls a year later.

At around this time Oswald claimed that Horace did not regard him too highly. On the other hand it is quite obvious that Horace was extremely fond of him and greatly admired his "Kid" brother. Nevertheless, this is what Oswald wrote:

> Horace professed to have a very poor opinion of my abilities as regards anything appertaining to engineering, in which I certainly had no training. He knew I was inventive since when I worked in his laboratory at Hove in 1899 I had designed an instrument for accurately timing a motor car's speed, and also an instrument for ascertaining the drift and direction of a balloon in relation to the ground. The principle employed in the very crude instrument I made was actually re-invented by someone and used by Imperial Airways Limited in 1928 as a drift indicator for aeroplanes.
>
> He [Horace] knew how we were left at Hove, when he had closed his laboratory. He knew that Eustace and I, starting with nothing, had managed to build up a flourishing balloon manufacturing business, and that it was through our connections with lighter-than-air vessels that he was now engaged in designing and building aircraft, and we were now equal partners with him. He was no doubt somewhat amazed at what we had done, but he never commented on it.

Oswald also said that

> at the stage of aeroplane design, neither he nor Eustace were able to give Horace any assistance – nor did he want any help from anyone. With his mathematical knowledge and his genius for engineering and inventing in many scientific spheres, he was completely at home immediately he turned his attention to aerodynamic flight.
>
> He was already known to some members of the Royal Aeronautical Society, by reason of his having given a joint lecture with Eustace before the Society on a means of reaching great altitudes by balloon in a pressure car. The inventor was Eustace, but Horace becoming for once interested in a balloon project, dealt with the mathematical problems and designed the car, and means of supplying it with fresh air at a pressure of ten pounds per square inch – a project carried into practice by Professor Picard and his company so many years later – in fact, the first pressurised cabin.

Perhaps Oswald misunderstood Horace's attitude because he still insisted on calling him "The Kid". The morning after Horace moved into Parsonage Farm, when Oswald met him at Mussel Manor and brought him over to the new workshops, the braziers had been lit and it was considerably more comfortable than Mussel Manor, although to be fair the caretaker had made Oswald a good breakfast. Horace was genuinely impressed with the way Oswald was organizing the building of the workshops, and now that the McClean order was under way, Horace was keen to be involved with future work in Leysdown rather than Battersea. It was interesting that, despite Oswald's talent, which he had already proven over and over again, he could not escape his own self-doubt that others, particularly Horace, might not believe in his abilities. Perhaps this accounts for the way he shrank from any form of, to use his own words, "puffing himself up". Tragically, Oswald's lack of self-esteem remained all his life despite his proven incredible inventive talents which gave so much to the evolution of our lives.

"This work is all commissioned by members of the Aero Club," explained Oswald.

"We need privacy," commented Horace at once, "and there is

not enough room here. And I need a place nearby to work on the drawings."

"We can begin to build the Shorts biplane No. 1 at Battersea," said Oswald. "There is room there and privacy, and you could use Mum's flat to do the drawings. I must contact Mr Griffith Brewer and ask him to take a lease on our behalf on this adjoining piece of land, so that we can set up here properly without any delays."

"Who owns it?" asked Horace.

"A man called Andrews; he is also a builder and no doubt would build us the sheds."

"Good!"

Before leaving for Battersea, they measured out the area they wanted and surveyed the possibility of a temporary aerodrome. They drove to Sheerness and parked outside the station, took the train to Victoria and made haste to Battersea to begin work on Frank McClean's order.

It was 22nd December. Horace moved into his mother's flat with his drawing board and graphics tools. The flat was filled with the acrid smell of cigar smoke while Horace chain smoked as he worked out the mathematical calculations for the aerodynamic workings of the machine. He worked solidly on the drawings, until the early hours of the morning. Too late to go home, he slept on his mother's sofa. At 5.30am, Oswald and Eustace, who had themselves worked late into the night at Battersea, crept quietly into the living room to take a look at Horace's work so far. Horace, a light sleeper, awoke immediately.

"I'll make you a cup of tea!" said Oswald quickly and left the room.

Horace rose from the sofa, and walked over to the dining room table where the drawings were laid out. Eustace followed him silently. It was the first time they had met since Horace had left Hove to go to Newcastle some seven years earlier. Horace smiled in greeting, but Eustace merely nodded his head in acknowledgement. Perhaps it was too early in the morning to make small talk thought Horace, wryly.

Eustace studied the drawings, nodding silently as he absorbed the design Horace was trying to achieve. He studied them for some minutes, trying to commit to memory the dimensions and calculations. Oswald returned with a steaming mug of tea which Horace took gratefully.

"Thanks Kid!" he said, rubbing his unshaven chin.

Oswald joined Eustace at the table and marvelled at his brother's exquisitely neat drawings. For a man with such large, clumsy-looking hands it was extraordinary how he could execute such delicate work.

"I haven't finished yet," said Horace.

When Marie Claire awoke in the morning she was not surprised to find that Horace had left. As always, she knew that she had to get on with life, sink or swim. She was a swimmer, but there were times when Horace frightened her. Although he could be the most charming, entertaining and affectionate of husbands and fathers, he was also given to violent mood swings of depression and anger. His energy was formidable, but very often when he had clearly been under strain through overwork he would lose his temper over the slightest thing, and suddenly erupt into frightening rages which were almost like fits.

Since making the decision to join his brothers, Horace was becoming more volatile and more taciturn. His excitement and impatience to get started was intense. She could deal with that, she would be as patient as she could be; it was the irrational rages which were unmanageable. Marie Claire had come to realize long ago what an unusual man she had married and what a brilliant mind he had. She could never hope to approach his level of intellect and there were many times when she did not understand his jokes, but she could look after him, cook his meals and try to make him eat properly, and try to make him rest when he was over-tired. She sighed; many a time she prayed that what she was doing was enough. But what more could she do?

She came downstairs in her overcoat to light the stove and to make breakfast. When they had returned from supper with the Rosewarnes, she was so tired and overcome by the journey and the whole emotional trauma of the move that she had barely noticed the house the night before. But now, as she descended the bare oak stairs, lightly running her fingertips down the polished banister rail, she began to appreciate her surroundings. There was a pile of chopped kindling and a bucket of coal in the alcove beside the chimney. She opened all the cupboard doors and was surprised to find everything she needed. She crossed the big flagstoned floor to the larder and found the milk and bread she knew Kate Rosewarne had left for her. What a relief she spoke French. Marie Claire found that when she was anxious, all the English she had learnt over the years disappeared from her head. She was anxious now – she knew she shouldn't be, she had moved around plenty of times with the family – but she really had thought that Northumberland was to be her final home. She doubted very much that Eastchurch was any more secure.

Parsonage Farm was an ancient farmhouse but it had been meticulously maintained over the years. The oak floors were in excellent condition and the panelling in the large, square hall and two downstairs rooms were all in good repair. The timber glowed in the candle light. The ceilings were quite low and beamed, and the doorways were small. All the rooms led off the hall. There was one very large room which could be either a dining room or a sitting room. It was a much bigger house than the one they had had in Northumberland, but she liked it and she liked the Rosewarnes, so today she would see if she liked Eastchurch.

When Leonard, Ruby and Francis with Gracie at his heels came downstairs for breakfast they found a crackling fire in the hall warming the whole house, and the stove in the kitchen boiling the kettle for tea. Marie Claire had never employed help in her life; she was used to hard work, and enjoyed cooking and housekeeping. Perhaps it was the French in her, she shrugged. She could hardly believe that Kate in that small house had two servants. What could they do? Extraordinaire!

The children had barely finished their breakfast when there came a knock at the front door. Leonard, with a cry of delight, put down his mug of tea and ran to open it. Dorothy and another girl of similar age stood on the doorstep, and Henry immediately bounded into the house to find his new friend, Gracie.

"This is my friend Florence," announced Dorothy. "Would you like to come and see Florence's house?" she asked bluntly. "It's got lots of ghosts," she added encouragingly. Marie Claire came to the door and invited the little girls in and quickly discovered to her relief and delight that Florence also spoke French.

"Where did you learn to speak French?" inquired Marie Claire.

"We go to a French convent in Sittingbourne. We have to speak French there all the time," replied Dorothy airily.

"Sittingbourne! That is a long journey every day," exclaimed Marie Claire.

"We're boarders. We come home at the weekends, mostly," said Dorothy.

Marie Claire was shocked that children so young were sent away to boarding school.

"Well," she said, "I am sure Leonard and Francis would be delighted to come and see your house, Florence, but Ruby has a cold and I think she should stay in."

"They can stay for lunch, if they like," announced Florence nonchalantly.

There was a succession of callers that morning. Kate Rosewarne offered Marie Claire the services of her young maid, Lotttie, which Marie Claire declined, but also offered to introduce her to the village bakery and butcher, which invitation Marie Claire gratefully accepted. Amy Cunningham, Florence's mother, called and said how pleased she was to meet Leonard and Francis and that they were welcome to stay at Shurland Hall that day for as long as they liked, they were no trouble. She invited the family to a Boxing Day lunch after the hunt at Stone Pitts Farm. Dulcie Dickson, the vicar's wife, called at tea time and invited the family to mince pies and hot chocolate on Christmas Eve.

Marie Claire and the children had no time to wonder about Horace that day and by the time they went to bed they were all exhausted. Especially Leonard and Francis, who were overwhelmed with the adventures they had had that day at Shurland Hall.

Even though Dorothy and Florence were girls, they were very good company. The following morning there was no sign of Horace but Charlie Rosewarne called early with a Christmas tree and helped the children and Marie Claire install it in the hall.

Oswald told Horace of their progress the previous day on Frank McClean's order and Horace digested everything Oswald had said. Eustace remained silent. Eventually they left for the Battersea workshops. Horace lit his first cigar of the day, and immediately began to revise the work he had done. All through the day he worked with barely a break.

His mother brought meals to him, but knew her son well enough not to interrupt. By the evening he was satisfied with his first draft sketches, and lay back exhausted on the sofa to wait for his brothers' return. He would stay in London one more night so that he could go to Knightsbridge in the morning and buy special Christmas presents for the family and a Stilton, candied fruit and vintage port from Harrods food hall. Marie Claire deserved a new hat. "The Kid" said he would like to come to Eastchurch for Christmas. He knew that Eustace would stay with Mum and help her entertain the family. Ernie, his younger brother by two years, was running the coal merchant's business now. He'd got nippers, and Alice, his sister, would probably come over with her brood. He knew that Leonard, Ruby and Francis had set their hearts on having their Uncle Oswald for Christmas.

CHAPTER FOURTEEN

A contract secured

Oswald made his excuses on Boxing Day and left Horace and his family that morning. The truth was that he really could not bear the thought of a lot of people on horses and a lot of blood-thirsty hounds chasing through the countryside after one poor old fox. The very idea of killing any animal was completely abhorrent to him. He was deeply touched to see the genuine disappointment on the children's faces when he announced that he must get back to London. He had had a wonderful time; the candlelit midnight service with carols on Christmas Eve and the packed church raising the roof with "O Come All Ye Faithful" had given him an unaccustomed lump in his throat. The warm hospitality of the Rosewarne family, and of Vicar Dickson and his wife, had all made him feel so welcome. He thought how fortunate Horace was to come to live in Eastchurch and what an extraordinary place it was, given its ancient history and the many famous people who had not only visited the village over the centuries but had also lived there. Oswald had sensed a strange air of excitement about the atmosphere, as if it were on the brink of something remarkable. Perhaps he was being fanciful.

Neither Marie Claire, Ruby nor Kate had any wish to follow the hunt either, so Charlie and Horace, with Leonard, Francis and Dorothy perched in the "dickie" at the back, piled into the Gordon Bennett. The children screeched in mock terror every time they went over a bump, however small. There was a sparkle in the air, and although it was cold, the sun shone and made the fields shine as if they had been specifically cleaned for the occasion. As they approached Shurland Hall, crowds of people from nearby villages were making their way across the field leading to the main gates, which opened inwards, revealing the courtyard where some of the huntsmen were already assembled. Frank Ingleton was the Master this season, and, resplendent in scarlet, he was now sampling the stirrup cup and looked as if he had been doing so for some time. He saw Charlie and gave him a shout in greeting. Many of the farmers from around the island enjoyed the hunt and had come that day to support a tradition that had survived for over fifty years. This was the first time in all that time that they had visiting gentry. There had been reports in the newspapers recently, describing machines which could fly, and there were rumours that some of the gentry here were those very flying pioneers who were experimenting with these "aeroplanes". Some rumours even said that they were coming to set up an airfield in Leysdown. The word had spread like wildfire, and the villagers came from far and wide to catch a glimpse of the inside of Shurland Hall and some of its distinguished guests.

Amy was mounted on her small black mare, and for once was riding side saddle, dressed in a very fetching riding habit and feathered hat. Florence was beside her on a small fat piebald pony. Dorothy, Leonard and Francis disappeared into the middle of the assembled hounds, and the huntsmen shouted at them to move away. Frank Ingleton sounded his huntsman's horn, and the riders whooped and jostled to leave through the gated archway. When they emerged, the horses pawed the ground impatiently, and then suddenly they were off across the fields towards Minster. With the hounds baying, and the hooves pounding, the riders called out and whistled to each other. The followers jumped into their motor cars and pony traps and set out along the winding lanes. Wrapped in thick rugs warding off the sharp wind, laden with leftovers from their Christmas Day dinners, they were prepared to follow the hounds to the kill.

Snow heralded the New Year in 1909, and many were confined to their houses, but Horace had managed to leave the Isle of Sheppey and head for Battersea before the Kingsferry bridge was closed. Oswald and Eustace were overloaded with balloon work, but like Charles Rolls it was a vicious circle; they needed the cash flow. Every morning Oswald and Eustace left before dawn for their Battersea workshops and worked late into the night. It had now become a matter of extreme urgency that their Leysdown project with the new sheds was finished as soon as possible, for there was not enough room at Battersea to build fixed-wing aeroplanes. Eustace and Oswald's reputation was now so solid that even the Army had commissioned them to make balloons, but still the Government remained sceptical as to the value of powered flight and refused to provide any funds for the development of aircraft.

News of Wilbur's seventy-seven-mile flight in December at Le Mans, winning him the Michelin prize, had spread like wildfire. Suddenly, experiments were being carried out by European pioneers to emulate the Wrights' success. Frank McClean and Griffith Brewer decided to visit the War Office themselves to try to make an appointment to see a minister but they were fobbed off with a junior official. Furious at this rejection, Charlie Rolls asked his father for help, but Lord Llangattock only received shallow excuses about priorities of budget. The fact was that the British pioneers were ignored and had to continue to use their own resources. Meanwhile, Charlie Hands, aware of their dilemma, reported in the *Daily Mail* that although the German Government had increased their budget by an alarming amount the British pioneers were once again being ignored. Even the public were becoming anxious at the lack of interest by the Government and letters began to appear in *The Times* expressing surprise that Wilbur Wright's achievement in Le Mans was not attracting the attention it should from the Government and the War Office, when there was now a very real threat from Continental Europe.

After reading one such letter in his club one morning in January, Griffith Brewer made the decision to travel to France that very day and see the Wright brothers face to face. Before leaving, he called in on the Royal Astronomical Society in Burlington House, and as luck would have it found Frank McClean there in the observatory. Frank had already seen the letter.

"Frank," said Griffith, "do you think we could get a syndicate together to manufacture aeroplanes ourselves?" Frank McClean did not hesitate.

"Of course we can. Leave that to myself and Charlie Rolls. We've already agreed that that is the way forward. You go and see the Wright brothers and the very best of luck. We are not going to be defeated by some petty, bourgeois, narrow-minded bunch of idiots in this Government. Telegraph me as soon as you have news, old boy. Oh, and give my regards to Wilbur and Orville. Tell them we can supply the very best manufacturer they are ever likely to find, here, in Kent."

So, taking the same route which Oswald and Eustace had used at the end of September, he arrived at Pau two days later. The warm sun of the South of France revitalised Griffith Brewer after the bitter winter weather in Britain.

Wilbur and Orville were just settling down to an aperitif before an early dinner at the Chateau de Bayonne when Griffith Brewer strode into the drawing room to see them. They were astonished, but very pleased at this unexpected visit.

"Why," said Wilbur as both brothers rose to shake Griffith Brewer's hand, "this is so fortuitous. We have been talking for days on how we should proceed. We are at a crossroads here, and we so welcome your opinion on the way forward. You see, despite everything we have done and everything we have achieved, we cannot convince either our Government or your Government to commit themselves to any contract with us. We can see for ourselves the awesome interest with the public now. That was not the case a few months ago but now . . . why, soon we shall be overtaken if we are not very careful."

"That is why I have come," responded Griffith Brewer. "Your flight at Camp d'Avours and the Michelin prize at Le Mans has galvanized flying pioneers into ever greater ambitions. I do earnestly implore you to consider contracting your designs to an independent manufacturer who will not be concerned with funding from any Government. If you rely on any Government funding your designs, who knows what restrictions you will be tied to."

"Well now, this sure is interesting. You know, I believe Orville and I could go down that road, providing customers order directly from us."

". . . and we supply them directly from the manufacturers," intervened Orville.

For a bishop's sons, they were remarkably sharp when it came to business, smiled Griffith to himself.

There was a silence as the three men considered the potential deal. Not so good for the Shorts financially, mused Griffith Brewer. This would give us a really good financial deal, thought Orville and Wilbur at the same time.

". . . and do you think you know of a suitable manufacturer who would agree with those terms?" asked Orville doubtfully.

"I certainly do, and indeed you met two of these men last September!" replied Griffith.

"Eustace and Oswald Short? Are you sure?" asked Wilbur.

"Yes, I am," said Griffith firmly. "They have just been joined by their elder brother, Horace, and he is without doubt the best and the most inventive engineer in Britain today. You could not pick a better partnership, believe me."

"But who would be wealthy enough to put orders our way?" inquired Wilbur despairingly.

"Leave that to me," responded Griffith Brewer. "I can guarantee you now that I can secure you your order." Griffith was almost one hundred per cent sure that the Aero Club would fill the order with very little persuasion.

The head waiter appeared discreetly at Wilbur Wright's elbow.

"Dinner is served. Monsieur and I have taken the liberty of seating Monsieur Brewer at your table."

"That is very kind, thank you so much," said Wilbur quietly.

The three men talked all the way through dinner of how their plans could be executed, and by the end of the meal it was agreed that Horace should come and visit them in Pau as soon as possible. They in turn would pay a return visit to see the Shorts. Griffith excused himself after coffee. He climbed the wide staircase steadily, resisting the urge to rush up two steps at a time in excitement. In some ways he would have liked to have stayed for another few days to soak up the warm sunshine, but he knew that time was too valuable to waste on the South of France's inviting climate.

When the train pulled in to the Gare de Lyon in Paris, Griffith had an hour to spare.

He headed straight for the telegraph office and sent a telegram to Frank McClean: "Object achieved. Shorts on board. Immediate orders required for first contract. GB."

When Griffith arrived back at Victoria Station it was nine o'clock in the morning. He went straight to his club in Piccadilly, bathed, shaved and changed, and lost no time in making his way to Battersea. There was a lot to discuss.

Horace was just finishing his breakfast at Evie's Café where he had demolished an enormous plate of fried egg, bacon, black pudding, fried bread and mushrooms washed down with a strong mug of tea. He flirted mercilessly with the waitress, who responded with fits of giggles and gave him an extra fried egg, claiming, "I'm only trying to keep that big gob of yours shut!"

As Horace left the café and made his way to the Battersea workshops he could see out of the corner of his eye a young man lurking inside a doorway, and as he approached the young man stepped forward.

"Excuse me, sir," the young man said, removing his hat to reveal a head of carrot red hair. "I'm from the *Daily Star* and I wonder if you would mind telling me something about your recent partnership with your brothers."

"Yes, I would mind," bellowed Horace at the unfortunate young man, who had to retreat a few steps, "and if I catch you snooping around here again" – Horace's voice rose to a pitch where the whole street could hear – "I'll whip your hide off!" He opened the gate and slammed it in the young man's face.

Horace, puce in the face, stumped across the cobbled yard to Oswald's tiny office and immediately saw that they had a visitor, Griffith Brewer. All three men were desperately trying to control their laughter, even Eustace. Horace softened, realizing how funny he must have sounded.

"That blighter was a reporter!" he exclaimed.

"From the *Daily Star*?" asked Oswald.

"How should I know. He'd got carrot hair," grumbled Horace.

"Don't worry, he's harmless and I know he's on our side; he's one of Charlie Hands' students. Charlie Hands of the *Daily Mail* is very much on our side," said Oswald.

"Humph!" grunted Horace.

Griffith held his hand out before there could be any more discussion about reporters. Horace shook his hand, relaxed and smiled.

"Good news!" said Griffith. "The Wright brothers have agreed to give the Short brothers the opportunity to act as their manufacturers, and they are expecting you in Pau within the next week or so, Horace!"

"Details of contract?" inquired Horace sharply.

"Charles Rolls and I will work it out in the next few days, and Frank McClean will put together a group of buyers."

"I have yet to meet the Honourable Charles Rolls. I look forward to that. We must all meet at Leysdown before I go," responded Horace. Horace remembered the reporter outside the gates. Where on earth did these blokes get their sixth sense from, to chase a story barely before it happened?

Frank McClean was at his home, Rusthall House, Tunbridge Wells, when he received the telegram from Griffith Brewer, which had been forwarded from Burlington House, and he understood immediately that it was down to him to secure orders. He was busy writing up his trip to China. Actually, he hated writing reports, and was very pleased to have an excuse to search out Charles Rolls and John Moore-Brabazon. His father had been a great intellect and he sorely missed his advice. He would have written this report in no time. Frank senior had died of pneumonia in 1904 in Brussels at the age of sixty-seven after returning from one of his expeditions. He had been a Trinity College Cambridge scholar with an honorary degree of LLD from the University of Glasgow, a Fellow of the Royal Society, a Fellow of the Royal Astronomical Society and a member of the Institute of Civil Engineers. Frank had by no means inherited his late father's intellect but was neither jealous nor envious of him, merely proud. Frank had trained as a civil engineer, but had not distinguished himself particularly. But Frank had a strong, stable personality with an abundance of common sense and a huge sense of humour, and people loved to be with him. The one thing he had inherited from his father was his enormous generosity and his curiosity in anything innovative. His father's extraordinary adventures in astronomy had captured the imagination of the young Frank McClean at an early age, and in due course he had been able to advance some of his father's great work in that field.

There was no doubt that Frank McClean senior had had a similar intellect to Horace Short, although Frank McClean senior had all the advantages of a well-to-do middle-class background, unlike Horace who had left school at fourteen. It would have been an interesting meeting had their paths ever crossed. Frank was yet to get to know Horace properly but he had already identified the fact that he was an extraordinary person, brilliant, complex, but hugely entertaining.

One could never really say that dear Charlie Rolls was very

entertaining, mused Frank wryly. But Charles Rolls was the real visionary among the pioneers. He wasn't afraid of public opinion, and he was completely single-minded in his ambitions. His engineering degree at Cambridge he put to practical use, much to the derision of his own circle of friends. He was enterprising, daring, courageous and determined, all the most admirable qualities for a national hero. He was already a household name, pioneering the use of motor cars as the new method of transport. Without Charles Rolls, Henry Royce would have sunk into obscurity. He possessed the rare gift of being able to spot talent. Royce was extremely lucky to meet Charles Rolls with his knack for exploiting, convincing and selling to potential customers. No one could charm like Charlie, possessing dashing good looks, but he was nonetheless an enigma. Although he could be unbelievably rude and sarcastic, he attracted large crowds wherever he went, but he was basically a loner. Poor Vera Hedges Butler, now married to Iltid Nicholl, still followed him faithfully from one county to another, from one race meeting to another, even from one country to another. He was sure that Charlie wasn't blind to her devotion to him but there was nothing that could break social convention without causing awful, harmful scandal. Unfortunately, Charlie did possess a mean streak running through his veins which was almost embarrassing. It was well known that when it came to tipping, whether it be a waiter or a porter, he had always run out of change. Perhaps his father had kept him short all his life, although now through his own hard work and enterprise he was rich in his own right. Anyway, thought Frank, he would have to dig deep now, but he had no doubts that he would. Charlie's competitive spirit would never be held back by his financial idiosyncrasies.

John Moore-Brabazon was still only twenty-five years of age, more or less the same age as Oswald Short, but totally different in personality. John's wealthy aristocratic background gave him a boldness and self-assurance which put him at an advantage when faced with opposition or competition. Oswald, on the other hand, thought Frank, was shy, talented and modest beyond belief and had none of Brab's characteristics, but would achieve success despite this. One hoped that it would be recognized. Perhaps he was being unfair to conclude that Brab succeeded because of his background, because he did, after all, have the quality which mattered most when it came to experimenting at flying, namely courage. He had no doubts that Brab would want to buy one of the Shorts Wright aeroplanes.

Frank McClean, Griffith Brewer, Charles Rolls and John Moore-Brabazon drove down to Leysdown in one car. Charlie was the chauffeur. They found that the Short brothers had already arrived and were busy instructing Mr Andrews on how they wanted their own sheds constructed on their own piece of land. They were all surprised that the Shorts had already purchased some land.

Frank was taken aback that the Short plans were so advanced. The Aero Club had only one shed at the moment, which was totally inadequate for the expansion about to take place. He made the instant decision to purchase the golf course and land at Shellness and build at least a dozen flying sheds, so that they would have not only all the workshops and storage space they needed, but a proper airfield.

"Mr Andrews," said Frank genially, puffing his pipe, which had been extinguished by the wind and customary winter drizzle. "Could we go to Sittingbourne this afternoon do you think, and draw up a contract with my lawyers for the lease to purchase some of your land?"

Columns of pound signs leapt before Mr Andrews' eyes as he saw instant wealth coming his way, and almost without hesitation he replied, "Would three o'clock be suitable?"

"Splendid!" laughed Frank, reading Mr Andrews' mind accurately. "Let us adjourn for lunch!" They climbed into their cars and drove the short distance to Mussel Manor, where lunch had been prepared by the housekeeper. The table in what was the main room was laid for eight people. Frank turned to Griffith questioningly.

"I've invited a Captain Dunne; he is a member of the Aero Club, but he is with the Blair Athol. He is not due for another half an hour so we have time to discuss private matters between ourselves before he arrives."

"He is a fine designer! The Army have great faith in him. He has been to see us at Battersea and is extraordinarily knowledgeable despite his young age," said Oswald at the ripe old age of twenty-five!

Charles Rolls had been silent until then. "Good!" he said shortly.

"So," said Brab, "if you are going to have storage sheds built here then presumably I could keep my Voisin biplane here? I was going to keep it at Brooklands."

"Of course," replied Frank. "That is the whole idea. All the Aero Club may keep their machines here, and they can be repaired and serviced and so on and so forth, by Oswald or Eustace or Horace and their men. At the same time the manufacturing will carry on beside the flying activities."

"We shall be well away from prying eyes, and the interfering press, which is the great advantage of this place," said Horace.

"Indeed!" agreed Charlie Rolls. "All the while we are building and experimenting, the last thing we want is hordes of curious critics. But we shall succeed, and the Government will have to listen to us in the end."

Charlie Rolls and Horace weighed each other up and decided mutually that they respected each other's gifts and motivations.

"Of course they will," retorted Brab arrogantly. "Look at them: incompetent, unintelligent with short-term vision that is close to blindness."

Eustace and Oswald had never heard anyone speak so harshly of their Government before and were quite shocked. Frank and Griffith merely laughed at the outburst.

"When one of us flies across the Channel, that will be the moment when they will sit up and realize we are no longer safe from invaders as a nation," said Frank.

Charlie turned to Horace. "It is excellent that you are going to Pau and will meet Orville and Wilbur Wright for yourself." He gave one of his rare smiles. "You will find them cannier than any Scotsman when it comes to business."

"So Griffith tells me," laughed Horace, relishing the challenge of his forthcoming trip.

CHAPTER FIFTEEN

John Moore-Brabazon makes the first British flight in April 1909

Horace's trip to Pau took him just over a week. The journey was tedious and took longer than usual, due to the appalling weather in northern France. He found that both Wilbur and Orville were charming, but behind the charm was a steely determination to extract the maximum amount of money possible. He had been highly amused that, as the sons of a bishop, charity did not enter into any of their negotiations. Horace quickly disclaimed his responsibility for financial details and concentrated purely on the technical design. This immediately put them at ease. He was amazed that they had no technical drawings of any sort. Starting from scratch, he acquired all the information he needed, and made several draft sketches. The brothers were deeply impressed at Horace's evident skill and were mindful of Griffith Brewer's claim that he was the best inventive engineer in Britain.

Wilbur was six years older than Horace, who was thirty-six years old in 1908, and five years older than his brother Orville. In contrast to his quiet, intellectual brother, Orville was a real live wire, full of energy but not given much to intellectual studies, although possessing a truly inventive mind. Orville had been a champion cyclist, just like Charles Rolls. The two brothers complemented one another, and were obviously very fond of each other. They were always able to work things out without having any major disagreements. Horace was very impressed with their relationship, if not a little envious, recognizing perhaps that this was what made them such a solid and reliable partnership. He resolved that he would try to imitate their example and make more effort with Oswald and Eustace. Before Horace left, he knew that the contract was in the bag, and the brothers had promised him that they would visit England in May when the weather was warmer.

He arrived home tired and irritable, with a crashing headache. When he sat down to make some adjustments to his drawings barely before he took his coat off, Marie Claire protested that he should just for one evening relax and talk to the family about his trip. Unreasonably, he lost his temper. Ruby and Francis ran out of the kitchen, while Leonard bowed his head, afraid to look his father in the eye lest his father see the tears welling in the corners of his eyes.

"This work of yours is going to destroy us," cried Marie Claire.

Horace was silent. He rose from the table and went to Marie Claire and put an arm around her and beckoned to Leonard, who rose and went to him. He put his arms around them both.

"I'm sorry," he said remorsefully. He called out to Ruby and Francis. "Let's start the evening again, shall we!" For the rest of the evening he was a warm and loving husband and father.

After the children went to bed Marie Claire thought that it would be a good time to tell Horace that Leonard was not doing well at school. Half the time he absconded; she was sure he was bored, and the village school was not stretching him. Ruby needed some new clothes and Francis was becoming very rude. The sink was blocked in the kitchen, the coal merchant was not delivering the right quantity of coal and she needed more blankets.

"I'm tired, Marie Claire; you will have to sort out these domestic problems yourself. Go and see Kate Rosewarne; she will be able to help you." He could feel his temper rising. Marie Claire had ruined what had been, after a bad beginning, which he had to admit was his fault, a warm, enjoyable family evening. He stifled an outburst.

"Goodnight, Marie Claire," he said and went upstairs to bed, leaving her to dampen the stove down and shut off the gas lamps, a job he usually did when he was at home.

The following morning Horace left before the children were up, although they were awake. They heard the front door shut quietly, their father's footsteps crossing the gravel yard, the motor car being fired up and disappearing up Church Road. Now that Horace was rested, he put his domestic worries behind him. Marie Claire, with Kate Rosewarne's help, would find a new school for Leonard, and if it was going to cost money it did not matter.

Horace could barely see as the wind and rain drove against the windscreen so that he almost missed the turning to Shellness. Dim lights washed over the ground outside his new workshop, which Andrews, as good as his word, had finished. He could see "the Kid" standing in the entrance waiting for him. "Good lad," said Horace to himself, smiling.

The day after Horace had left for Pau, Oswald decided to move down to Leysdown and oversee the builders' progress on their work shed himself. It was difficult making that decision, because Oswald knew they had a full order book for balloons. Eustace had employed four extra men to help complete the order for aerial navigation for the Committee of Imperial Defence, and there were other orders waiting to be completed. Bent over his enormous workbench, wearing only a shirt and waistcoat despite the cold, his sleeves rolled up, Eustace was surprised when he saw Oswald standing before him in an overcoat, with a suitcase, two portmanteaux and several artists' folders.

"Eustace," Oswald addressed him quietly, "I've decided to move down to Leysdown and supervise the building of the work shed. We have got to have it finished by the time Horace returns."

Eustace sighed, shaking his head. "I think this order is more important, Oz; it is assured cash which we do not have much of just now. We're miles off finishing Mr McClean's order, you can see that. We need the money. Supposing the Wright brothers don't come through in the end? Then we're finished," he said with finality, his brow furrowed with anxiety.

"We have to be optimistic, Eustace, and look at the bigger picture. Horace will come back with the order, Griffith Brewer assured us of that. We are not in a position to waste a single day. Come on! We've been in tighter spots before," said Oswald, desperately wanting Eustace to see his point of view.

Eustace sighed again, shook his head, looked up with a half smile and said, "Go on then, I'll manage, but we've got to keep this part of our company going at full strength, Oz, whatever happens."

Oswald laid his hand affectionately on Eustace's bent back. "Thanks," he said briefly.

Little did Horace know the lengths to which Oswald had gone to make sure the shed was finished for his return. The weather had been atrocious all that week, but Andrews' men, working against the clock, achieved a remarkable job. Whether or not it was because Oswald himself helped the men with the construction, it was difficult to tell. Not far away, work had begun on the sheds for the Aero Club. Members were clamouring to move to Leysdown. Charles Rolls had come three times already that week to see progress.

"I've a good mind to roll my sleeves up like yourself," he said, not smiling, and leaving Oswald wondering whether he was serious or not.

Staying at Mussel Manor certainly helped achieve his goal of getting the shed built on time but it was a spooky place to stay on his own, no doubt about it. Oswald didn't believe in ghosts but it was so dark at night that it was almost dangerous to go outside in case you got lost and could never find the place again. The odd hooting of owls or sea birds and the baaing of sheep sent shivers down Oswald's spine, despite his usual calm, unshakeable personality.

Oswald was always first on site, for like Horace he needed very little sleep. He heard the sound of Horace's motor car approaching and went to meet him. He could tell by the broad smile on his brother's face that everything had gone well in France.

"Welcome back, Horace. What do you think?" said Oswald as he swept his arm out to show the shed.

Horace strode into the work shed. "Excellent!" he boomed. "Well done, Kid!"

Oswald winced. Horace walked over to one of the work benches and flipped open his notebook, which contained four sketches of their new Shorts Wright aeroplane, all drawn from different angles.

Griffith Brewer was the first to arrive that day and was astonished that the Shorts' shed was up and ready for work. Charlie Rolls arrived shortly after him. He examined Horace's sketches in silence, and finally announced that he would like two aeroplanes, one now and one later. Moore-Brabazon was next to arrive confirming his order for a pusher biplane "complete without engine". Frank McClean followed him. He was deeply impressed with the Shorts' shed and layout generally. Much to Horace and Oswald's amazement, he also ordered a Wrights design on top of the Shorts biplane already in production. It was a generous gesture and not lost on Horace and Oswald. They discussed the design and production in depth until they were interrupted by the sudden approach of a galloping horse drawing up suddenly outside, a loud snorting and neighing and a "Wooah there!"

Cecil Grace removed his rain cape and felt cowboy hat, flung them over a wooden packing box, and stood grinning in the entrance of the shed, exhilarated by the ride and the anticipation of the Short Wright aeroplanes. His energy and enthusiasm always brought a smile to everyone around him. He had an ability to make those around him feel good and forget whatever gloomy thoughts they might have at the time. Whatever Cecil tackled, whether it was riding a horse at breakneck speed, racing a car or flying an aeroplane, which he had every confidence he would be doing in the not too distant future, he would give his all and enjoy every minute of it. Cecil was not a boastful person, not at all; he just enjoyed life and wanted everyone around him to do the same.

"Hey, you guys," he laughed, "leave one for Percy and me!"

"That is it!" announced Griffith Brewer grinning broadly. "The remaining two aeroplanes are spoken for by two other members of the Aero Club, Alec Ogilvie and Maurice Egerton."

They paid £1,000 each, out of which the Wright brothers took £400 profit for each aeroplane while the Shorts were paid £200 for their labour and materials for each one. The engines were provided by the Wrights themselves, costing around £400 each, and according to Oswald had been built by Marre of France, versions of their original Dayton engine. The contract was officially signed in March 1909.

In March a magazine called *Flight* was launched, and published an article virtually letting all the cats out of the bag at once.

> Many people will doubtless have wondered what the Wright brothers were going to do with regard to England and the introduction of their Flyers into this country and few indeed could have known that

the well-known Aeronautical Engineers Messrs. Short Bros. have, for some time, been engaged on their construction. It was not that there was really any need for secrecy in the matter but merely because Messrs. Short Brothers themselves have been so very anxious to avoid any suggestion of "puffing" themselves by means of the Wright name that the information has hitherto been regarded as strictly private and it is only because Mr Griffith Brewer, who is acting as Patent Agent for the Wrights in the country, considered it proper to make the matter public, that it has become known at the present stage. It is not difficult to guess that applicants for the right to build the famous Flyers in this country have been sufficiently numerous and importunate in their petitions and it is by no means difficult for those who know the Short brothers to understand why those other brothers in France should have given them the first chance. They have a factory and are used to working with that care and precision on which depends the safety of all who risk their lives in the air, moreover, since the eldest brother has come down from the North – where he has lately been associated with the Hon. C.A. Parsons in certain experimental work – the firm represents a combination of theory and practice such as is dear to the Wrights' heart.

He has entrusted them with the building of six machines, which represent a money value of £8,400 and, for the moment, that is all that has been decided on, for the Wrights are essentially cautious people who know how much depends on their forethought in matters relating to flight, similarly with the Short brothers, who, because they wished to show that they could satisfy these scrupulous Americans before they talked about it, sought to keep their work behind scenes.

Presently, when the first Flyer is finished, there will be trials on the Aero Club's private ground at Shellbeach, where, perhaps, Wright himself may one day show us "how it's done".

Horace and Oswald exchanged looks of quiet delight but remained silent. They stood back as the pioneers, one by one, examined Horace's notebook and the draft sketches he had made. The atmosphere at that point was tense with excitement, all of them generating a force of energy of confident success within their grasp. Their absolute faith in the Shorts' ability to produce an aeroplane to fulfil their ambitions was steadfast but Horace and Oswald met the challenge with equal expectation of success. They knew what was at stake, and they knew they were as well equipped as any other manufacturer. They were ready for the enormous responsibility of launching a machine that was always going to face unknown dangers and put the lives of its pilots at permanent risk. Horace and Oswald observed these men, among the first flying pioneers in Great Britain, and not one demonstrated an ounce of fear or doubt.

The rain eased off in the afternoon and Frank McClean was able to inspect the sheds under construction. He pointed out to his contractors how quickly the Shorts' shed had been built and challenged them to do the same. Defiantly, they accepted the challenge.

Suddenly it dawned on John Moore-Brabazon where the airfield was to be marked out. He let out a howl of dismay when he realized that three perfectly good fairways were going to be ruined, and his games of golf were going to be badly affected.

"Brab, for goodness' sake!" snapped Charlie. "You can play golf anywhere you like on a very much better golf course than this."

"Sacrilege, old boy!" his friend moaned in mock pain.

"Come on! Come and have afternoon tea with me at the vicarage with Dulcie Dickson."

Brab was about to refuse, shuddering at the thought of a provincial vicarage tea.

"Her brother is Lord Coleridge, a good chap, friend of my father's," said Charlie, watching Brab wryly, as Brab went through a cocktail of indecision.

"Ah yes . . . love to!" responded Brab at last, unable to resist a wave of curiosity as to how someone like Dulcie Dickson had integrated into the village society of Eastchurch.

They said their farewells to Horace and Oswald. Frank McClean shook both men warmly by the hand. "Well done!" he said. "Well done!"

During the following weeks Oswald and Horace worked night and day at Leysdown. Skilled craftsmen heard of the project and came from miles around to get the chance to work on a Wright Brothers aeroplane. Few of them knew of the Shorts, but they had by now all heard of the Wright brothers' flying feats.

Poor Charlie Hands got short shrift from Horace whenever he popped up, which was frequently. Horace was determined to play down the Wrights' involvement, because he knew ultimately that he would design his own aircraft. It was a rather arrogant self-opinionated ambition, because Oswald was also a talented designer, despite his gap in formal education, and over the next few weeks Horace finally began to realize just how accomplished "The Kid" was.

Charlie Hands was "fit to be tied". Finally these blighters had pulled the big-time contract and here he was sitting on the greatest story of the century unable to write a single word about it. Until one of them got into one of the machines and flew it, no one was going to pay a blind bit of notice, and these blokes didn't deserve more scepticism and derision. He'd watched Charles Rolls sweep down that little lane to Leysdown time after time, and he did not dare get a story out about him either. What could he say? He knew that everyone around was sworn to secrecy about his presence in the area, so far be it for him to blow the whistle. It achieved nothing, and he'd probably get the sack. He would just have to wait and see. "Patience will out!" Charlie kept telling himself. Week after week he visited Leysdown, skulking around and keeping well out of the way of Horace Short. He watched the sheds being built for other pioneers being completed in next to no time.

"You here again, Charlie!" Frank McClean would call out good-humouredly.

Charlie sighed and went back to the Castle pub, ordered a beer and sat down to write an article about the site only.

> The new works which the Short Brothers have built on the flying ground will be a revelation to anyone who looks on the construction of flying machines as being merely a hobby for wealthy amateurs. The firm has about forty men hard at work, and is steadily increasing the number. The main shop is wide and high enough to permit the erection of two complete machines at a time, as well as leaving room for putting together the sections of several others, and for the general carpentering and joinery work of manufacture. Adjoining the main shop is a spacious drawing office. Also built on to the main shop is a small but complete smith's shop, where the metal joints and lugs are

> produced. At the time of my visit to the flying ground, the part of the works devoted to construction was occupied by two machines, namely, Mr Moore-Brabazon's Voisin biplane, which was under repair after a slight accident during trials, and Mr Frank McClean's Short biplane, which is approaching completion. There seems to be every likelihood of a regular flying colony springing up at Shellness.

Charlie Hands knew his article would probably be hidden amongst the back pages somewhere. He would just have to wait until someone actually flew for a sustained period. He ordered another beer and lit a cigarette. He would just have to be patient and pick up gossip where he could, and this was probably the best place. Horace often met Charles Rosewarne at the Castle, and he knew that his kids were friends with Horace's and he knew that Charles Rolls had approached Rosewarne to build him something on Stone Pitts Farm, away from the Leysdown site. Very often the Shorts' workmen came in for a pint at lunchtime and shared a chat with the locals; if he bided his time, sooner or later, he would hear exactly what is going on.

There was much more concern with the general public now, for people were beginning to be aware of the French and German Government's interest in flying machines, and at last realized the corresponding lack of interest by the British Government. Blindly and stubbornly, the Government were waiting for results which they were not prepared to help along. Ridiculously, they wanted someone to fly across the Channel before they stepped forward to help and acknowledge the contribution of so few. The fact that it could be a Frenchman or a German who could snatch this prize from under the noses of their own flying pioneers did not seem to matter a jot. Even the general public could see that the threat of invasion was now a reality. Many asked how their Government could be made to take the English pioneers seriously. How could they be made to see sense? In the end, would it all be too late? Charlie Hands lit another cigarette, feeling the frustration himself. He desperately wanted to help to bring these pioneers' efforts to the attention of the public, and force the Government through public opinion to do something. He knew he couldn't without a proper story, and nor could any other reporter, not until something was worth reporting.

Finally, Charlie Hands' patience paid off because, one day in the middle of April, as he sat in his usual place, Oswald and Horace's workmen came and sat at a nearby table and in hushed tones discussed the Wright brothers' visit to Leysdown at the beginning of May. Charlie practically bit his cigarette in half and choked on his beer, all at the same time. He sat hardly able to breathe, shrinking into the seat so as not to be noticed. Eventually, they got up and left. Charlie literally ran down Station Road to catch the Sheppey Light to head back to London. It was certainly his lucky day, because he bumped into Griffith Brewer, catching the same train.

"Ah, Charlie!" Griffith greeted him.

Charlie waited for Griffith to tell him the news of the Wright brothers' impending visit but instead he said, "I'm sure you will be interested to know that Mr Moore-Brabazon is going to fly his Voisin biplane which he has brought from France, the *Bird of Passage*, at the end of April. It has been modified by the Short Brothers and it can be said that it is mainly the first English model to attempt a sustained flight. You may be interested to know, however, that the propeller was made in France by a firm that makes lavatory seats." Griffith looked Charlie straight in the eye without a flicker of any facial expression. "We have every expectation the tests will be successful. You should come!"

"Thank you very much," said Charlie, dying to ask about the firm that made lavatory seats, but he waited instead for Griffith Brewer to tell him about the Wright brothers' visit. He said nothing. Never mind, it would be easy to fabricate where he had heard the news.

"Let me know if you have any other interesting news," he said artfully.

Griffith Brewer nodded vaguely, and it seemed to Charlie he deliberately avoided any more conversation by climbing into a carriage further down the train.

The weather was perfect for the weekend's flying experiment. It was a beautiful mild sunny April day and the air was fresh and clear. Yellow daffodils decorated the hedgerows bordering the twisty lanes down to Leysdown, the sea was a dark navy blue and the fields towards the marshes were a riot of different shades of green. Members of the Aero Club and their friends and wives had all gathered for the occasion. Charlie was surprised at the number of people who had come. He quickly sought out Oswald, who was standing on the edge of the crowd talking to his brother Eustace. Horace, he was happy to see, was at the forefront of the crowd with Charlie Rosewarne, surrounded by children. Cecil Grace appeared to be accompanied by a small, dark-haired lady and his brother Percy. He recognized Frank McClean and one or two others from the Aero Club. He presumed that many of the other onlookers were the Short employees and local villagers. The gathering had all the atmosphere of a kind of carnival.

"Good turnout, Oswald!" Charlie remarked, nodding to Eustace at the same time. "I don't see Mr Charles Rolls here?"

"No, as a matter of fact he couldn't make it," replied Oswald hesitantly.

"He's in Derby; something urgent came up concerning his car business," said Griffith Brewer smoothly.

"Oh," said Charlie Hands as nonchalantly as he could. "I thought he'd gone to meet Mr Wilbur and Mr Orville Wright at Dover."

Griffith Brewer grinned. "Always one step ahead, Charlie!"

"You can't keep something like that secret!" replied Charlie quickly.

"No, you are right!" he agreed. "But actually it is true, he is in Derby, he's collecting his new Silver Ghost which he aims to drive to Dover to collect the Wrights and in fact I'm going to accompany him. They are coming here on the second of May, after the weekend. I advise you to bring your camera, Charlie; it will be some occasion, I can tell you."

"Thank you, Mr Brewer," said Charlie seriously. "I will be discreet, but I will make sure that people in the highest circles know of this visit, and are made aware of its significance and importance."

Oswald and Eustace remained silent, unwilling to contribute anything to this important conversation, so crucial to their future. Charlie Hands was simply never going to understand why all three of the Short brothers were so determined on keeping everything so quiet. There they were, the first manufacturers of aeroplanes in the world, and they wanted to keep it a secret. They should be famous, they should be household names, they should be signing autographs and have their pictures in the papers. Charlie felt exasperated with them. Modesty was one thing but this attitude would never get them the recognition they deserved. They should take a leaf out of Mr Moore-Brabazon's book; he knew exactly how to make the most of his talents. He was a superb showman, observed Charlie, laughing inwardly and hoping for the young man's sake that he would actually get the rather cumbersome machine off the ground and not kill himself in the process. Success or not, he was going to arrange that the *Daily Mail* would offer £1,000 for the first flier to complete a circular mile. He had no doubts that John Moore-Brabazon would take up the challenge.

The moment came and everyone held their breath as the ropes were relinquished and off he went, but only along the ground. The crowd shouted out their encouragement just as if the machine was actually a living being. Finally they were rewarded and John Moore-Brabazon launched his biplane three times that afternoon. A hundred-and-fifty-yard hop, a two-hundred-yard hop and a five-hundred-yard flight. Each time he rose from the ground the crowd cheered and clapped, excited and astonished all at once. Even Eustace and Oswald shared the crowd's emotions, although it was Horace who led the cheers.

Brab was not in the least bit worried about his engine and the ability to eventually get into the air; what really worried him was his ability to land. He could see the ground was riddled with dykes and ditches and when on the final occasion he managed to get a longer flight he was suddenly struck by a strong gust of wind which tipped the whole aircraft sideways. So he heaved on the rudder control rather too much and broke it. Suddenly, unbeknown to the awestruck crowd, who were unaware that anything had gone wrong, everything was beyond his control, and all he could rely on was a bit of luck. He glided down rather too heavily and struck the tip of the left-hand wing. The machine was so delicate that the impact shuddered through the machine and the wires and struts all snapped so that the engine came hurtling through the air from behind, missing him by inches and burying itself in the ground. Typically, he clambered out as if the accident was just a slight mishap. It was only later that Charlie was able to see the extent of the damage and realized what a narrow shave he had had. It made Charlie think hard. This was just the beginning. How many would be so lucky? How many would sacrifice their lives in the months and years to come?

When the flying machine finally came to a halt, the children ran down what had been the first and second fairway of the golf course, followed by the crowds. Although modest in number, they were enthusiastic in their admiration and congratulations to the somewhat dazed but triumphant pilot as he lay prone on the ground, his face being licked by a dog. Charlie was as excited as anyone in the crowd, not just for John Moore-Brabazon but for the Aero Club as a whole. Now he had something to say that should make people sit up and take notice. This day, 30th April, 1909, Moore-Brabazon was the first Englishman to take flight in Great Britain.

This was how Brab described his extraordinary achievement in his own words:

> 'This flight, which I suppose may justifiably be termed 'historic', was not a very long one nor was it very high – perhaps fifty feet. My main anxiety to begin with was not on account of the engine, which pulled extremely well, but on account of the aerodrome: it was so full of dykes and ditches that landing could be very dangerous if not made in precisely the right place.
>
> But I had greater anxieties to follow.
>
> While in the air I was suddenly struck by a powerful side gust of wind. The whole aircraft of course tipped sideways and the impact was so sudden that I had no time to do anything but try to correct my position in the air by heaving on the rudder control. This I did, but in my anxiety I proved a little over zealous and suddenly discovered to my horror that the control to the rudder had broken. So there I was in mid-air and absolutely unable to control my machine.
>
> There was nothing to be done but glide down – canted at an angle of about thirty-five degrees – and this I did. The whole thing was beyond my control. I remember seeing the pattern of ditches and dykes spreading out below me. Only a few seconds before, I had been imagining that by summoning all the skill I possessed, and with a lot of good luck added, I might be able to land without damage. Now I had only good luck to rely on.
>
> The machine pitched forward and sideways even more, the ground span below me, and suddenly I struck the ground heavily on the tip of the left-hand wing. The impact shuddered through the machine, wires and struts snapped viciously, and with the shock of

impact the engine left its moorings and came hurtling through the air from behind me – missing me by inches – and buried itself in the ground.

It was a startling and disconcerting experience, but through good fortune I wasn't personally damaged. A bit bumped and bruised and dazed, and pinned down by wires, but without a scratch, I recovered to find myself being licked by my two dogs, who had chased after me during the flight.

Brab then goes on to say something that most people might disagree with, that 'they were not in any way brave!'

I do want to stress something here about which there may be misapprehension: the idea that we were brave. We were very interested, and we may have been foolish, but we were not brave; for nobody had had any accidents and nobody had been killed. Aviation took place chiefly on the ground, and aviation on the ground is not particularly dangerous. It is only when you get up that danger starts, and it was only just then starting. It would be allowing an error to creep in if I let it be understood that we were heroes.

That was Brab's opinion, but they were heroes of course. They defied governments, the press and public opinion to make the ambition of flight a reality.

At last Charlie Hands had something to write about, and with 2nd May looming, it seemed that he would get just the coverage he had been aiming for, for weeks.

CHAPTER SIXTEEN

The Wrights visit Mussel Manor

Griffith Brewer stood beside the console table at the bottom of the stairs in the large entrance hall at South Lodges. He heard running footsteps two floors above him and looked up to see Charlie Rolls peering over the banisters, closely resembling one of his Derbyshire mechanics, hair dishevelled and his striped, collarless shirt, even from where Griffith was standing, did not look clean. Charles leant over the rail.

"Apologies Griffith, I didn't get back from Derby until late last night, and I had to go into Conduit Street first thing this morning."

Griffith Brewer looked at his pocket watch again; it read five minutes past six in the morning. This was "first thing this morning". Charlie was working round the clock, determined to make a success of his motor car business, Rolls-Royce Cars, as he now called it, despite his total commitment to the Aero Club, the Short brothers, his own ambition of achievement in aeronautics, and ensuring that eventually the public and the Government would acknowledge the importance of the future of flight. His order for two aeroplanes was generous, and Griffith knew the Short brothers appreciated his generosity. Charlie's complex and volatile character, with a reputation for being spectacularly mean when it came to tips and parking fines and paying for restaurant meals, starving rather than pick up the bill for a meal, was quite at odds with this huge gesture. Charlie would order the cheapest item on the menu and would travel second class while his friends would travel first. Tommy Smith always paid his speeding fines, which Charlie would then tell him to get reimbursed from the company secretary. Porters on the station at Eastchurch and Leysdown always mysteriously disappeared on the few occasions when Charlie travelled by train, as they knew they would get nothing in the way of a tip. And yet he would happily slip a sovereign to a son or daughter of a friend if they would recite a poem or sing a song or simply perform a task for him. No one would ever get to the bottom of Charlie Rolls' complex character.

"Smith is putting fuel into the Silver Ghost," called Charlie and disappeared from sight. Griffith heard a door close and he prayed silently that Charlie's valet was making him change his clothes. Not more than three minutes later, Griffith heard a door open on the second floor, the immediate sound of running feet for a few seconds, then silence, running feet along the first-floor landing, then silence. Griffith looked expectantly towards the staircase, and in the next second watched Charlie Rolls' descent at breakneck speed down the banisters of the staircase which many members of the Royal Family and aristocracy had descended rather more sedately. Charlie landed lightly and elegantly on his feet, proving this was another of the accomplishments he enjoyed practising.

Griffith, examining Charlie's clothes, was astonished at the quick change his valet had achieved: starched white Eton collar, starched white shirt, waistcoat with pocket watch, immaculate suit and even the tie looked as though hours had been spent tying it. It was rumoured that Charlie's valet had been a vaudeville dresser well used to the art of quick changes, and it probably partly accounted for the fact that Charlie knew so many vaudeville songs and appeared to be word perfect in most of them, and loved performing them whenever he had the opportunity, usually to an open-mouthed audience who couldn't believe what they were witnessing. Was this really the Honourable Charles Stewart Rolls?

Charlie so regretted being unable to attend the momentous occasion of Brab's first flight, but it was impossible for him to be in two places at once. His obsession with flight, and the urgency that he, the Shorts and his friends should succeed and show the country how mistaken they were, had made him all too aware that to achieve this goal he had to sell cars, lots of them. The need was compelling and immediate, and it was a viable means to an end, because the cost of their experiments was immense. When he heard that Brab had done so much damage to his aeroplane, he wondered how on earth he was going to pay for the repairs. So far as he was concerned, he knew that being self-effacing was a luxury he could not afford, and he must use every means he could, and never mind his mother's harsh criticism that it was middle class and common to demonstrate one's wares. He was prepared to do whatever it took to make his marketing strategy effective. False pride and snobbery was a luxury, and something he had never understood, nor indeed wished to. It had been a source of friction between him and his mother all his life. He was barely on speaking terms with her and his hurt had gone deep. Creative invention, that was the heart of progress. How he admired the Wright brothers and the Short brothers' dedication and creativity. Charlie sincerely believed that he had a duty and a mission to bring success to these gifted men, whatever the price.

When Charlie wasn't at Leysdown, he was either in Derby or in Conduit Street attempting to find ways of enticing the rich into his showroom, and getting the maximum publicity wherever he went. If his mother were ashamed of the way he touted for business, it was "hard lines". He had come to terms with her incomprehension of anything he had ever done, but it made him more defiant and determined and more ambitious. On the other hand, he knew that his father was inordinately proud of him, and his heart warmed when he thought of the concern he saw etched in the lines of his father's face whenever there was a problem. He acknowledged that his father's concern was not just for financial reasons, but over the backlash of cruel words from friends, the public, the Government and the press, when it was clear that his son was risking his very life for the sake of progress. Where Charlie's relationship with his mother was cold, his father was his mentor, his friend and rock. Nevertheless, he did not expect his father to fund him indefinitely. He had been more than generous, from the very start of his enthusiasm and ambition to launching his first passion, the motor car. Now Charlie had moved on to an even greater challenge and still he knew he had his father's moral support; he knew how his father had to put up with jeers and patronizing disbelief, even from his own friends. More than anything, Charlie wanted to repay his father in proving that these heavier-than-air machines were essential to the defence of the realm, and the breaking of new barriers and exploration into unknown territory in the world. It was going to take money, lots and lots of it, and it was down to him to provide what finance he could through his own efforts.

He had already informed Lord Northcliffe that he was going to meet Wilbur and Orville Wright on 2nd May in his Silver Ghost, and expressed the hope that photographers would be at the port of Dover to record the occasion, paying particular attention to the fact that the brothers were being met by a Rolls-Royce Silver Ghost!

Charlie Hands took a gamble and decided to avoid the melee of reporters and photographers at the port of Dover, and headed straight for Leysdown. Griffith Brewer had told Hands that Charlie Rolls would get there just as fast as he could after collecting the Wright brothers from Dover, hopefully, without being caught for speeding.

It was a cold day. The fine weather of the weekend had disappeared, and grey clouds covered the sky as if winter had temporarily returned. The pioneers were all on edge, and were surreptitiously watching the lane and listening out for the purr of the Silver Ghost. Oswald had changed out of his work clothes and into an elegant dark grey suit. He would not have been out of place in any gentlemen's club in London. His brother, Eustace, had changed from his work clothes and looked equally elegant.

It was difficult to spot the similarity with their brother Horace, who in contrast looked more like a country farmer ready for the market, wearing a lumpy tweed jacket and thick twill shirt, with his tie askew and heavy workman's boots. He seemed quite out of place with the rest of the group and yet was perfectly at home with them. Oswald and Eustace were clearly furious with him for "letting the side down", looking so untidy, and Oswald was fussed that Horace would call him "Kid" in front of everyone. With his pipe clamped to his mouth, Horace impatiently patrolled Mussel Manor as he waited for Charlie Rolls to arrive.

Frank McClean had ordered lunch from Simpson's in the Strand, and waiters in white aprons were to be seen through the windows of the ground floor of Mussel Manor preparing lunch. Charlie Hands knew that he would not be invited and that he would have to be content with cheese and pickled onions in The Billet at Eastchurch. Still, he thought, cheering up, he could hunt out Charlie Rosewarn; he was a close chum of Horace's, and he might drop some useful information.

Sitting on the wall outside All Saints Church in Eastchurch were Dorothy, Florence, Leonard, Ruby and Francis, who had been told by Charles Rosewarne that some very, very, important and famous people would be passing that way around lunchtime. They had waited there patiently for nearly an hour, when to their huge excitement, an enormous motor car approached them down the little high street from the direction of Sheerness. They jumped off the wall and waved at the motor car's passengers but, to their surprise, the motor car stopped, and the driver got out.

"Good day to you, ladies and gentlemen," exclaimed the driver. The children giggled, and Dorothy, who was definitely the boldest and most forward, replied:

"Good morning sir, we've been waiting for you."

"Ah," said Charlie, gesturing to his passengers. "Let me introduce you to Mr Wilbur Wright and Mr Orville Wright, who are the most

Orville Wright, C.S. Rolls (driving), Wilbur Wright and Griffith Brewer visiting Leysdown in May 1909. (Courtesy of the Royal Aero Club)

famous aviators in the world. This is my friend Mr Griffith Brewer and I am Mr Charles Rolls."

"How do you do," said Dorothy. "This is my friend Florence Cunningham, and these are my friends, Leonard, Ruby and Francis Short. They moved here at Christmas time," Dorothy explained. "My friend Florence sings," Dorothy added as an afterthought.

"Why, how delighted we are to meet you!" exclaimed Wilbur and Orville, lifting their hats.

"It is very nice to make your acquaintance. I hope we shall meet again soon," said Charlie, climbing back into the Silver Ghost.

Charlie Rolls never ceased to amaze Griffith; he appeared to be more at home with children and artisans than he ever was with his own kind.

"Goodbye," said Charlie and his passengers, waving to the children.

"Goodbye," called the children.

Finally, those waiting anxiously at Mussel Manor heard the engine of the Silver Ghost in the distance together with the distinctive sound of the klaxon, and suddenly there they were, Wilbur and Orville Wright in Leysdown, the Isle of Sheppey. Charlie Hands stepped boldly forward and asked them if, before they went inside for lunch, they would assemble for a photograph in front of Mussel Manor to commemorate this historic occasion. And before they had time to disperse, he swiftly organized the group, hustling Mr Andrews, Oswald, Horace and Eustace Short beside Frank McClean, Griffith Brewer, Mr Hedges Butler, Dr W.J.S. Lockyer and Mr Warwick Wright, with Orville and Wilbur Wright seated in the front between John Moore-Brabazon and Charlie Rolls. He quickly threw the black cloth over his head, and clicked the shutter. It was all done in a matter of minutes and Charlie Hands had scooped what was arguably the shot of the century.

It was a pity that Alec Ogilvie, Maurice Egerton and either Percy or Cecil Grace had been unable to get to Leysdown that day. Charlie Hands would have liked to have had a photograph of all those who had signed the contract with the Short Brothers, the first manufacturers of aeroplanes in the world. That would also have been a photograph worth having.

The most famous 'line-up' in early aviation taken during the Wrights' visit to Leysdown on 6th May 1909.
Back row from left to right: Mr. J.D.F. Andrews; Oswald, Horace and Eustace Short; Frank McClean; Griffith Brewer; Frank Hedges Butler; Dr. W.J.S. Lockyer and Warwick Wright.
Seated left to right: J.T.C. Moore-Brabazon; Wilbur Wright; Orville Wright and the Hon C.S. Rolls.
(Courtesy of Medway Archives)

The day turned out to be very successful. Wilbur and Orville thought the flying fields were perfect. They did not observe the hazardous dykes and ditches. But what took their breath away was the technical drawings Horace had made of their aeroplane. They examined everything in the work sheds and were full of praise and enthusiasm. Although Horace took over, Griffith Brewer noticed how both Wilbur and Orville drew Eustace and Oswald into their conversation. They were well aware that their elder brother, genius that he was, had joined their company, and not the other way around.

The shine was slightly taken off the day for Charlie Rolls when his good luck ran out down Bluebell Hill on his way to Rusthall House with Wilbur and Orville Wright, when the police, ever vigilant, caught Charlie speeding. History does not recall what he said to them but one can imagine. When a passing hansom cab called out "Old Iron", Charlie quickly retorted, "Take it home and eat it!"

A lavish dinner was also given during Wilbur and Orville's stay, in their honour, by Frank Hedges Butler, to which Vera, now married, was invited, and at which Charlie was invited to speak. Still smarting from the police trap, he told his audience that flying would undoubtedly become popular in the absence of dust, dirt, taxes and police traps. This was as usual a prophetic claim since at the time there had only been one short flight by John Moore-Brabazon, which had ended abruptly.

But Charlie was confident, knowing that this was just the beginning, and when the diners yawned discreetly, he went on to say that the dropping of bombs from airships was already a practical idea that could well be used in any future war. That was another truly prophetic claim: the Zeppelin onslaught was only five years away.

The following day Wilbur and Orville Wright had an appointment with Lord Haldane, the Minister for War, but all they received from him was a lukewarm handshake and lukewarm interest in the fact that six of their designs were about to be manufactured by the Short Brothers, who were to become the first manufacturers of aircraft in the world. It was a bitter rebuff to both sets of brothers.

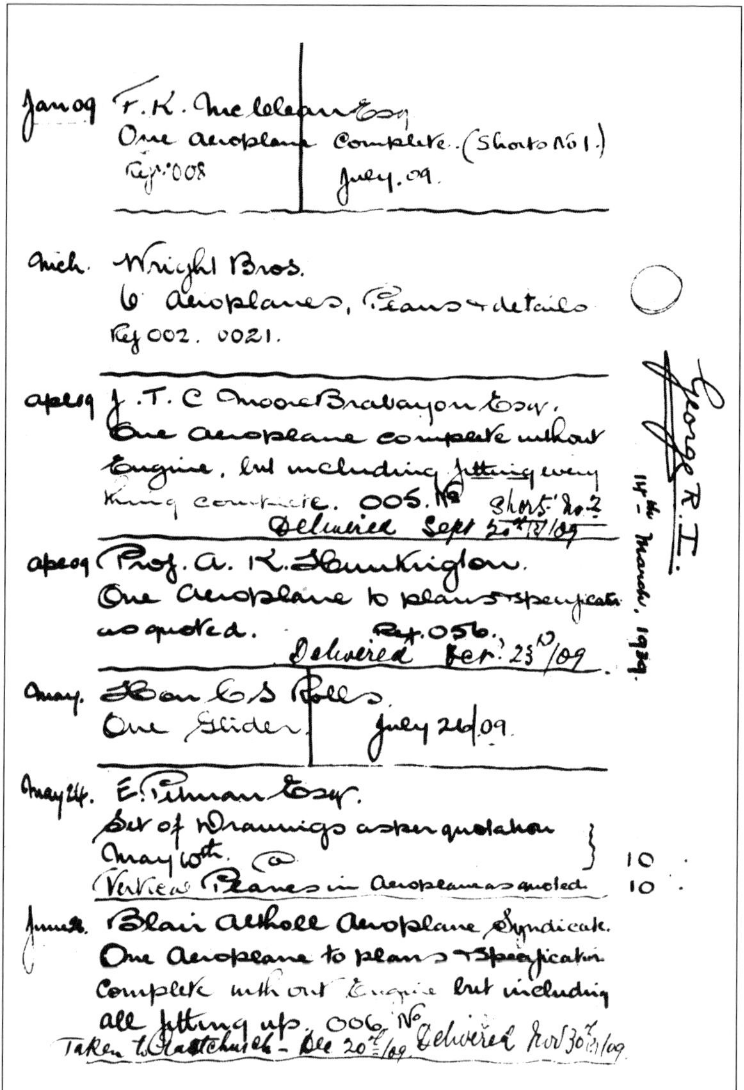
Jan 09 F.K. McClean Esq
One aeroplane complete. (Shorts No 1.)
Ref. 008 July. 09.

Mch. Wright Bros.
6 Aeroplanes, Plans & details
Ref 002. 0021.

Ap 09 J.T.C. Moore Brabazon Esq.
One aeroplane complete without
Engine, but including fitting every
thing complete. 005. No Shorts No 2
Delivered Sept 20th/09

Ap 09 Prof. A. K. Huntington.
One Aeroplane to plans & specification
as quoted. Ref. 056.
Delivered Feb. 23rd/09

May. Hon C S Rolls.
One Glider July 2nd/09.

May 24. E. Pitman Esq.
Set of Drawings as per quotation
May 10th. @ } 10 .
Vertical Planes in Aeroplane as quoted 10 .

June. Blair Atholl Aeroplane Syndicate.
One aeroplane to plans & Specification
Complete without Engine but including
all fitting up. 006 No
Taken to Eastchurch - Dec 20th/09. Delivered Nov 30th/09

George R.I. 14th March, 1909.

First page of Short Order Book 1909: Jan 09: F.K. McClean, Esq.;
March: Wright Bros; April 09: J.T.C. Moore-Brabazon, Esq.;
April 09: Prof. A Huntington; May: Hon C.S. Rolls; May 24: E. Pitman, Esq.;
June: Blair Athol Aeroplane Syndicate; Aug: Hon C.S. Rolls

CHAPTER SEVENTEEN

Life at Mussel Manor: Charlie learns to fly and Brab flies the circular mile

Oswald now lived at Mussel Manor almost entirely. Since his move to the Isle of Sheppey his eyes had been opened to country life and the rare beauty of the wild marshes and extraordinary birdlife. He had become squeamish about eating meat, confronted by all the baby lambs in the fields and marshes all around him. He loved to watch the families of baby rabbits bounding in and out of the hedges near Mussel Manor, and found that he could not face eating his mother's rabbit pie any more.

Oswald was joined that spring by Captain John William Dunne, a most inventive designer, who, having studied the science of aerodynamics and flight, built test models of aeroplanes which were tailless and had swept wings, giving the machine stability. As a serving officer he had been sponsored to test fly his machines. Unfortunately there was a only limited success and the Army withdrew their funding which was how Captain Dunne, anxious to continue his experiments, came to Leysdown, where he was welcomed warmly by Oswald. As with so many pioneering ideas, his designs were met with scepticism but he ultimately proved to be one of the most successful forerunners of aircraft designers.

Oswald writes that:

> Dunne was a good companion. He was a crack shot with a rifle, with which he used to shoot wild duck on the marshes. I accompanied him with a .45 Webley revolver, firmly determined not

> to shoot at a living thing, but we had good practice at a target. He taught me some ju-jitsu holds and we fenced with cardboard tubes made to hold drawings.

The other spasmodic visitor to Mussel Manor was John Moore-Brabazon, who was always very lively company and interesting to talk to. His particular hobby at the time was colour photography. It was at its earliest stages, when you viewed the photographs under a microscope, and what you saw was a myriad of tiny spheres in a multitude of colours.

Despite the terrible, wet, late spring and early summer of 1909 the pioneers continued their experiments at Leysdown with extraordinary zeal, determined to ignore the attitude of the Government. They were undeterred by the appalling conditions and the struggles endured in retrieving their aircraft from unseen bogs and ditches. Charlie was determined to learn to fly but there was no way he was going to choose Leysdown for his experiments. He went in search of Rosewarne, and called one evening at his house, which backed on to Church Road and was the entrance to Charles Rosewarne's building yard. Charlie walked through the yard and found the small gate which led through an apple orchard up to the back door.

Dorothy answered the door and recognized Charlie immediately.

"Good evening, Mr Rolls," said Dorothy, quite assured.

"Good evening, Dorothy," said Charlie, recognizing Dorothy as one of the children he had stopped to meet with the Wright brothers. "Is your father at home by any chance?"

"Yes," said Dorothy. "Please come in; I will go and find him." She disappeared through a door off the room he had entered, which he presumed was the dining room, and observed the large table, too big for the size of the room, laden with a high tea. Just the sort of food he liked: boiled eggs, herring rollmops, plates of bread and butter. One could barely move round the table, what with the enormous sideboard and an upright piano.

"Ah, Mr Rolls," boomed Charles Rosewarne, appearing suddenly, "would you like a cup of tea? What can I do for you?"

"Thank you, Mr Rosewarne, I would indeed like a cup of tea." Charlie sat down at the table with Charles Rosewarne and explained his predicament, how he was anxious to get on and teach himself to fly but needed a spot away from Leysdown. Leysdown had become too difficult a place to experiment. He needed somewhere right away from the marshes, preferably a hill from which he could launch his aeroplane. Rosewarne told him that he thought he knew just the place, five minutes from there, at Standford Hill, just below Parsonage Farm, the house the Shorts had rented.

"Excellent," said Charlie as, to Rosewarne's great amusement, he tucked into the high tea on the table as if he hadn't eaten for weeks. "Would it be possible to park my car in your yard from time to time while I am experimenting?"

"Certainly!" replied Rosewarne.

The following day Charlie found Standford Hill, which seemed to be exactly the sort of hill on which he could practise to his heart's content. He was determined to learn to fly as soon as possible. The Government's demand that someone should fly across the Channel before they would step in and invest in the pioneers was an overwhelming challenge.

However, out of the blue, on 25th July, 1909 Louis Blériot shocked the pioneers by achieving this challenge and won the *Daily Mail* prize of £1,000 for his historic flight.

Predictably the Government still did not react. Charlie knew it was only a matter of time before a return flight would be made, and a German could leap in before a British pioneer had a chance. The Government were truly blind and the only way was to bring the pioneers' efforts to the attention of the public, who would, he hoped, force the Government to do something.

Charlie set about making preparations for the challenge and made copious notes on the experiments he undertook. As usual, he was methodical and extremely careful. His glider, built by the Shorts, used a starting rail, similar to the powered Wrights, but on the very first day of his tests there was very little wind and, what with a series of derailments, the tests came to a grinding halt. The following morning he was up at 4.00am.

"Four attempts with railway down hill," he records, 'but would not rise. Then put rail to top, at first attempt machine lifted nose for the first time for 5 or 6 ft, remaining in air about 2 seconds. It veered to right owing to rail not facing wind, and cracked right skid and broke a wire."

Experiments were held up, first for the necessary repairs, and then by wind conditions – too little, then too much. It was not until six o'clock that he succeeded in making a straight glide of eighty-five feet. His summary of the day's labours was terse and self-critical: "I am making landings still too much on back end of skids – though much better."

Even at the beginning of these experiments, Charlie knew where the mistake lay, which was the Wrights' dependence on the starting rail and pylon, which meant that the machine could not take off again unless these were in place. It was a very tedious business, as described by Tommy Smith, Charlie's long-suffering mechanic:

> We had to arrange a monorail 100 feet in length, facing the wind, which often changed before the job was completed – then we mounted the machine on a cross bar straddled across the rail; at the rear of the rail was a pylon, and we had to haul weights of about half a ton to the top of this. A rope was made fast to the weights over some pulleys; then the rope was taken about halfway down the rail over a pulley wheel, and from here to the front of the aeroplane. When all was ready for a start, Mr Rolls would release a catch, and the weights began to drop, which, with the rope attached to the machine naturally assisted it along the rails. One can imagine what happened the first time he attempted to make a flight. Having got everything ready, the engine was started, and at the final let-go the machine careered along the rail, with what seemed at the moment a terrific speed. When it reached the end of the rail, it suddenly shot straight up into the air, and a second afterwards crashed down on its rail with a tremendous bang. We found Mr Rolls calmly getting out of the mess, fortunately unhurt. He said: "This is my first lesson." I asked what happened, and he replied: "I am afraid I gave the elevator too much lift, but better luck next time."

It was not until the following summer of 1910 that this contraption was done away with and replaced with wheeled undercarriages.

Meanwhile Brab, who was very keen to win the *Daily Mail* circular mile flight prize, was busily repairing and rebuilding his *Bird of Passage* but, to use his own words,

> I was anxious to win the Daily Mail flight, so I concentrated on the new one. Short No. 2, which was the machine built for me, had no undercarriage with wheels, only skids; I think the technique of launching, to be appreciated, should be understood.
>
> First of all, imagine a pylon about eighteen feet high with a pulley on the top – rather like the top of a pit shaft to look at. A rope ran over the pulley and down the hollow of the pylon and had a half ton weight attached to it. From the foot of the pylon projected a single rail about sixty to eighty feet long, and on this rail was placed a sort of roller skate with a sleeper on it. And on the sleeper was poised the plane. The rope attached to the weight inside the rail, and along the end of the rail and back to the aeroplane – to which it was attached by a catch mechanism. You got into the machine and started the engine (how simple this sounds!) and, having got it up to full revs, you pulled the trigger – whereupon you were pulled by the weight and pushed by the propeller, so that when you had travelled on your skate to the end of the rail you found that you were doing about forty miles an hour and you had to do something about it by pulling your elevator. After that if you got into the air, well it was your job to get on with it.
>
> The snag with this pylon launch affair was that you had to set the rail so that it pointed directly into the wind, and invariably when you had got it all set the wind changed and you had to rearrange the rail. When I say wind, do not imagine that the wind was such as we would designate by that name today. It was a recognized thing in early aviation that if when smoking a cigarette the smoke from it went away from you, then you knew that it was virtually a gale in which it was too dangerous to fly.
>
> Well, when you got into the air there was nobody to tell you what to do, how to balance fore and aft or how to make a turn. In fact on taking turns we never banked – that was a very revolutionary introduction in turning which developed later. If, after all these years, there is any recollection of that circular flight of a mile on the first

English machine, it is of the turn at the end of the half mile. To this day I remember I turned perfectly flat with no bank at all, how the wind changed and instead of coming straight in my face came very much from the side, which was a new experience. However, it didn't seem to have any bad effect from the point of view of lift or balance, and I didn't slip to the ground. It all seems fairly amusing now, but at the time the control of a machine such as mine, which was extremely unstable in every possible way, was in essence an adventure into the unknown. One knew that such and such ought to happen as a result of certain movements one made at the controls, but whether they actually happened was a very different matter. And if things didn't happen as theory laid down, then one had to try not only to keep aloft but also to discover the reasons for the unexpected.

What usually occurred in our launchings from the pylon was that when the machine came off the rail it fell off on to the sleeper and broke the skids – meaning more repair bills. That was one of the bugbears, the frightful repair bills. Every time you went out something was smashed and had to be put right.

When the machine was ready, I gave notice to the Daily Mail that I would like this flight that I contemplated making observed by them, as well as by, naturally, The Royal Aero Club, of which Commander Perrin was then secretary. Flying, as I have said, was only possible in almost still air, with the result that every morning we got the machine out before daybreak and waited for dawn. The Daily Mail sent down to observe this flight a most extraordinary character named Charlie Hands. I was very honoured to have Charlie Hands as observer. He had vast experience of the world, was one of the most distinguished correspondents of the Daily Mail and was much revered by everybody in the press world. A more human and more lovable man I have never met. He used to go into the local pubs and immediately was at his ease and everybody adored him. I count my contact with him as one of the privileges of my life. Poor Charlie later went blind, but I shall cherish for a long time the memory of a dear companion in very difficult times.

Every morning he got up with me at about 3 o'clock, and for a solid fortnight we went back to our little cottage, Mussel Manor, at

Leysdown, the wind having got up before the sun. Instead of going to bed, from four o'clock until nine he insisted on playing piquet, and although I had not played before and he had to teach me, by the time we parted I was quite a proficient player.

Eventually the day arrived when things were all right. I was shot off the rail, balanced myself for half a mile in the most remarkable way, got round the post at the other end and flew back and landed just where I had started. I don't suppose I was at any time more than twenty feet from the ground, but I never actually touched, though I was jumping up and down the whole time. Balancing fore and aft was exceedingly critical, for this machine had no stabilising control whatsoever, and was very tricky in pitch.

I should very much like to put one of the first-class pilots up in this machine today and see how he would get on with it. It must be remembered that one was unable to practise on it in any way, so that to have taken it up and flown it for the circular tour of a mile was a feat of some dexterity. For this I won the £1,000 given by the Daily Mail . . ."

CHAPTER EIGHTEEN

Move to Eastchurch

By the beginning of October the nights were drawing in and dawn was getting later but the activity amongst the aeronauts was becoming more intense than ever. Numbers had increased considerably, not only in the workshops, where there were now more than forty workmen, but also the numbers of pioneers. They were all up at dawn, waiting to see what the weather had in store; and if the wind got up before the sun and their cigarette smoke blew ahead of them, they were unable to experiment, but work continued until the last light. Crews had to work until late after dark retrieving the small aeroplanes from across the fields, very often with the assistance of Frank Ingleton's horses. It had been a wet summer, the ground had unseen boggy ditches, and even though they had burnt some of the ditches so that the aeronauts could see from the air where to land, there were some expensive disasters. So far, no one had flown any great distance despite the fierce competition amongst themselves.

Charlie's friendship with the American, Cecil Grace had become closer over recent months They were always competitive, but in a sympathetic way, each very considerate to the other's needs, and especially considerate to their own mechanics. Cecil was popular wherever he went but Charlie, although loved by his mechanics, remained aloof to strangers and acquaintances, especially those who had no interest in motor cars or aeroplanes. Both Charlie and Cecil were unhappy with the flying fields at Leysdown, where there were too many unseen hazards.

Horace, Oswald and Eustace Short, Frank McClean and Charlie Rolls were being driven at a pace they could not control. Their lives were no longer their own, and they were being pushed by the speed of the evolution of their own endeavours. Charlie had taken to regularly parking his Silver Ghost in Charles Rosewarne's building yard, which was not far from Parsonage Farm and Standford Hill.

Charlie, through sheer willpower and tenacity, had finally taught himself to fly. He took valuable time off away from his business, practising regularly at Standford Hill which overlooked Stone Pitts Farm, the distant marshes and the River Swale. Day after day he practised and perfected the skill he needed to succeed in claiming the valuable prizes which were now emerging from newspapers and rich enthusiasts. Rosewarne built flying sheds for Charlie at Stone Pitts, so that he was close to Standford Hill and his new practice ground. As a result of this move, Charlie came to the difficult and momentous conclusion that Eastchurch was a much more suitable place to operate than Leysdown.

At the end of a working day it had became a ritual to meet for high tea at Rosebank, Charles Rosewarne's bungalow, with Charlie Rolls, Horace Short, Frank McClean, Cecil Grace and whoever had been with them that day at Leysdown. They would sit around the ebony table in the tiny dining room and discuss their progress. It was at one of these informal meetings that a decision was made to move to Eastchurch. They could no longer continue at Leysdown, where the unseen hazards had become so costly. They would move, lock, stock and barrel, to Eastchurch, which would be a much more suitable place from which to operate.

Frank asked Rosewarne whether he thought the Cunninghams would be willing to let or sell some of their land around Shurland Hall. Rosewarne thought they would be delighted since he believed that Mr Cunningham had recently lost a lot of money through a bad banking investment. However, he knew for a fact that the property belonged to Mrs Cunningham through a trust made by her father, which was just as well, otherwise the bank could have taken the property. Current laws decreed that a wife's property actually belonged to her husband, but the trust had circumvented this law.

"Not surprising", twinkled Charles Rosewarne, "that she has joined the suffragette movement."

"Good for her," said Charlie Rolls swiftly. "I believe Vera Nicholl has also joined; we should introduce these ladies to each other, unless of course they have already met."

It was going to be a massively complicated move but, undaunted, Frank McClean set about the organization of setting up new workshops, purchasing land to build living quarters and renting four hundred acres from Shurland Hall for the flying fields. With his extraordinary organizational skills and his overwhelming generosity as the main benefactor, the move was achieved remarkably smoothly and swiftly.

One Sunday, during a church service at All Saints in Eastchurch, Charlie Rolls and his friend Cecil Grace were in the process of moving their aeroplanes from Leysdown, where they had been stored while the new sheds were being built at Eastchurch. Tommy Smith, Charlie's long-suffering mechanic, was due to help them load the precious machine on to a trailer, but was late. So, in a mood of sheer bravado, Charlie decided to fly the aeroplane to Eastchurch. Cecil immediately volunteered to accompany him. There was no sign of Tommy, so like two truant schoolboys they pulled the machine into the middle of the little runway, which had once been the second fairway of the golf course. They sat laughing on the grass as they released the ropes.

"Poor Mr Smith! He'll just die when he sees the empty shed. What will he think?" giggled Cecil.

"Serve him right for being late!" said Charlie, grinning. "I shall leave him a note! '*The Cage is Empty! The Bird has Flown!*'"

"Charlie, how cruel you are!" exclaimed Cecil and the two laughed all the more.

The service drew to a close. The church wardens, Kate Rosewarne, Mr Pearson and Mr Ingleton put their collection purses on the plate for the blessing and walked sedately up the aisle towards the altar. The Reverend Dickson made the sign of the cross and the wardens turned to walk back to their places to receive the final blessing. The Reverend Dickson turned back to face his congregation and gave his final address as Amy struck up a triumphant finale and the congregation and choir made their exit.

As the congregation emerged through the main door, and the choir, who had flung off their surplices, emerged from another door at the side of the church, the sound of the droning of a huge cloud of bees appeared to be flying their way. They all spread quickly across the churchyard and shaded their eyes to see what was coming.

Charlie and Cecil spotted the crowd.

"Yahooo!" shouted Cecil with his cowboy cry. "Perfect timing, old chap; look, they've just come out of church!"

Charlie, always a showman at heart, guided the small craft over the churchyard. He and Cecil waved to the astonished crowd below. The small aeroplane looked like a toy, the wings like those of a dragonfly, and it seemed unbelievable that it was able to fly with two men between its wings. They could see the Reverend Dickson still in his white cassock, the wind blowing it against his body so that it billowed out behind him like a sail caught in a storm. Charlie saw him making the sign of the cross as he swung round heading for Stone Pitts. Adrenalin flowed through both men like a drug; they felt invincible.

Charlie landed the Short Wright biplane at the bottom of Standford Hill on a level piece of ground in front of the new sheds. It was the end of October, but the weather was better now than it had been all summer, and in two short months Charlie had taught himself to fly. His ballooning experiences had taught him how to handle air currents and his motoring experience had taught him how to steer and manoeuvre. As soon as Cecil had his hands on his order from the Shorts, only a month ago he had also taught himself to fly. He had some way to go to catch Charlie and Brab, but he thought that he was already proving to be quite skilful. Actually, Horace thought privately, he was going to be more talented than either.

Even though it was Sunday, carpenters and tradesmen were hard at work erecting new sheds and transporting old ones from Leysdown to be rebuilt. As they came to a halt and jumped down, feeling both exhilarated and ridiculously pleased with themselves, they saw Horace emerging from one of the sheds.

"That's torn it, old boy!" said Cecil, grinning nervously despite himself.

Charlie had to acknowledge to himself that he felt sheepish. Only his old headmaster at Eton and Horace could make him feel like that, even though Horace was about his own age. Taking Cecil as a passenger two days before he was due to enter the Salomons Trophy was a risk he should not have taken. Charlie decided to take the bull by the horns. As a matter of fact Horace did look a bit like a bull, he thought, and just in time he took the smile off his face as he approached him,

"Good day to you, Short! Sorry to buzz in on you like this . . . I know it was a stupid risk and all that . . . but, er . . . Smith was late and I needed to get going . . ." – Horace scowled and growled something unintelligible, having been taken off guard – ". . . we left a note for him which said '*The Cage is Empty, the Bird has Flown*'!"

Horace suppressed a smile, ready to dress both men down, but suddenly threw back his head and laughed. He couldn't help seeing the funny side when the serious Mr Smith read the note.

"Phew," whistled Cecil, feeling like a small boy caught in the act of something wicked.

They retied the ropes and pulled the precious aeroplane into the empty waiting shed, not far from where they had landed.

"Hallo, old man!" came a shout behind them.

Brab stood silhouetted in the entrance to the shed, arms folded, and legs planted a stride apart. Charlie Rolls was still Brab's hero, even though time had moved on and he was able to compete with Charlie on the same terms.

"My dear fellow!" said Charlie, walking towards him holding out his hand, "Congratulations on your *Daily Mail* prize. How absolutely splendid," he said as he shook Brab's hand warmly.

"By Jove, the thousand pounds is most welcome. I should say so," preened Brab, looking like the cat who had stolen the cream. "Blinking expensive these repairs. I absolutely had to do that circular mile . . . do or die! Charlie Hands, the *Daily Mail* journalist, was absolutely splendid the way he set it up and so conscientious in making sure everything was properly organized while he patiently waited by my side until the right moment came to attempt the challenge. We are lucky to have him on our side, you know. All set for the Salomons, old boy?"

"Well, like you Brab, it's 'do or die'! I shall win it, never fear; I have to. We're all short of cash, but we'll manage, we have to help each other, be prudent and all that. Mr Short is right, we can't afford a lot of accidents, especially unnecessary ones."

As the aeronauts talked, Amy walked back from the church with Florence, up the slope to Shurland Hall. Florence was nearly twelve and getting quite grown up for her age. Their relationship had grown even closer over the past two or three years of shielding each other from

John Cunningham's bad temper and alcoholism. But more importantly, as Florence's obvious singing talent developed, mother and daughter spent hours practising together and analysing each other's work. Dorothy envied the relationship and realized that it was special. After all, her Papa was kind and caring, unlike Florence's who was mean and cruel.

"It was amazing, Mama," said Florence. "You should have seen them, except you were still playing the organ. It was just as if the music you were playing was a special accompaniment, you know, as if it was all part of a theatrical scene. I don't suppose they heard you, but we could; the organ music playing in some sort of harmony. Mr Rolls flew quite low and he and Mr Grace waved to us all, and then they were gone. There was a funny kind of silence, everyone was holding their breath and then we all cheered and clapped and you stopped playing at the moment they disappeared."

Amy shut her eyes briefly, wishing Cecil wasn't so foolhardy. One thing to be a daredevil on his horse, but quite another in one of those frail aeroplanes. What hope had he got if he fell out of the sky; it was hardly like falling off a horse. She felt her heart hammering with fear. When she was with Cecil it seemed that the sun always shone, he was so warm and understanding. He was such a wonderful companion, with a brilliant appreciation of music, all sorts of music, and a superb horseman. It was the wildness about him she loved and feared.

John Cunningham had returned from the City on the Friday evening with a streaming cold, behaving in a way which made her think he was bordering on insanity. Amy cancelled the lunch engagement they had with the Dicksons and decided on a quiet Sunday and to try to make her husband understand the scale of these pioneers' ambitions. Sadly his view was typical of most of the British public. She recalled the conversation her husband had had with Frank McClean when he had visited them to sign the lease on their land.

"Do you suppose this Government will ever commit themselves to your . . . er . . . experiments?" John Cunningham sounded cynical, but Frank McClean had ignored this and replied seriously:

"Do you know, it is an extraordinary thing, but they simply will not move until they feel we could be under the threat of invasion. By the time that happens it will be too late. Even Louis Blériot failed to get their attention when he flew across the Channel this July."

"Somebody flew across the Channel this July? How shocking!" John Cunningham sounded faintly bored.

"No, it was an incredible feat, and one which we were trying to achieve before anyone else. The only shocking thing about it is that the Government are still sitting on their hands, refusing to recognize a threat that is now real, and are leaving the responsibility entirely to private individuals courageous enough to persist, regardless of the physical and financial perils."

John Cunningham remained silent, unable to respond. Who were the idiots in this, he wondered callously.

Amy was well informed of the struggles of the fliers through Cecil, but when she had tried to tell John Cunningham he was always dismissive and uninterested. Perhaps it was pointless to try to convince him once again. She and Cecil were drawn together initially by Cecil's appreciation of music and then their love of riding. To Amy, Cecil had become a true friend and confidant and he was someone who made her laugh and lightened her life. She was not sure whether she was in love with him, and preferred not to think about that, but she was anxious that her husband should have no reason to suspect there was more than just a casual acquaintanceship. She glanced at him briefly and felt a wave of revulsion. The ravages of alcohol had made his face fleshy and his complexion an unhealthy, florid colour. He was a big man but he was overweight, he lumbered rather than walked, and when he sat down his thighs strained the fabric of his trousers. It was always an effort for him to get up from a chair.

Competition amongst the pioneers grew to a fever pitch as new inventions and discoveries were being made almost daily, and yet still the public were sceptical. Even the villagers of Eastchurch were not all that convinced. Tommy Smith and John Adams, Cecil Grace's mechanic, recall having a well-earned pint at the end of a long day in The Billet, and how the locals could not understand why "these mad blokes wasted so much time and money on such dangerous contraptions". To the world they were still the lunatic fringe of the engineering world and the playthings of the rich.

Of course Charlie was certainly not playing. He was striving to achieve success and as a result had to release some of his obligations with Rolls-Royce, which did not help his pocket. Even his natural parsimonious instincts did not help much. It was a vicious circle. His experiments with his aeroplanes, and the setbacks which occurred, were proving much more expensive than he had realized. Competition was increasing faster than he could have imagined. Each aviator wanted to be the first, although so far the Eastchurch aviators were still in the lead. Horace would always be one step ahead; and now they had their new flying fields thanks to Frank McClean's recent acquisition from Shurland Hall, everything was in place except for the finance. With the best will in the world Frank McClean could not continue to be their sole banker.

Charlie and Tommy Smith shuttled backwards and forwards from London to Sheppey by train. Charlie was always considerate to Tommy, and even John Adams remembered him as being one of the friendliest and most considerate of pilots, unlike his attitude towards many of his contemporaries, who found Charlie rude and cold. Tommy was a typical cockney, slightly built, with sharp features and a sharp wit to match which Charlie liked. He had married young, when he was nineteen years old, and had fathered two children by the time he was twenty-three. Charlie always inquired after Tommy's children.

As 1909 came to a close Charlie made his mark as a pilot by successfully flying for one and a half miles to win the Salomons Trophy. In fact Brab, who could easily have achieved this, sportingly agreed to suspend his attempt by a week to give Charlie a chance. The rigorous demands they made on themselves were also made on their mechanics and John Adams recalled:

> "There weren't any union rules. We were up at four o'clock every morning waiting for the weather. I was lucky – Mr Grace had a cottage, and my wife house-kept for the Grace brothers, myself and our rigger, so there was always something to eat before we went to work.

Until the aviators mastered their aeroplanes, the early days were the hardest, but Charlie, and indeed Cecil, took pleasure in attempting to conquer what Rolls termed "the recognized standard of impossibility". In those early days there were many, many "recognized standards of impossibility". Brab even took a piglet on a flight, to prove to a past sceptic that "pigs could indeed fly".

J.T.C. Moore-Brabazon proving that "Pigs might fly"

The Hon. C.S. Rolls at the controls of his aeroplane.
(Photo by courtesy of the Royal Aero Club)

CHAPTER NINETEEN

Charlie goes to Nice and makes a plan

The year of 1910 promised to be an exceptional one, and the calendar quickly began to fill with a series of meetings throughout Europe, from Biarritz to Berlin and to St Petersburg. The Royal Aero Club gave its first pilot's licence to John Moore-Brabazon, and the second to Charlie Rolls. Their certificates were dated 8th March, 1910. Cecil Grace was included in the first ten awarded.

Charlie owned three different aeroplanes in 1910. The Short Wright was passed on to Colonel Capper, which he and Smith delivered by road to Laffan's Plain at Farnborough, and Charlie offered, in due course, to help train Army pilots on the Short Wright. Sadly, this offer was never to be fulfilled. The second aeroplane was a Sommer, but for some reason he did not get on with it. The third aeroplane was a French-built Wright biplane. Just as Charlie's design and development contribution was made in the manufacture of his motor cars, so too was his input into one of his aeroplanes. This was reported in a flying magazine:

> This is not the Wright Bros' machine owned by Mr Rolls, but a species of Voisin type embodying a number of his own ideas. Mr Smith tells me that this aircraft was given the designation RPG (Rolls Power Glider) and was still under construction when its creator was killed.

Charlie already had ambitions to develop his own aircraft manufacturing company. His plans did not include the services of Henry Royce, who had already proved his total lack of interest in Charlie's flying achievements, and continued to complain that he should be more involved with Rolls-Royce Limited, despite Charlie's extraordinary sales record.

In April 1910 Nice held an international flying competition, which was the biggest event in the spring of that year and had the largest number of competitors, of whom Charlie was one. The authorities spent some £36,000 on it, an enormous sum, and more than the British Government was willing to spend on the whole National venture. The prizes were attractive, and some were as much as £8,000, but Charlie knew that the competition was very fierce. All the elite fliers of Europe would be taking part and most of them had already written him off, knowing how recently he had qualified for his pilot's licence, and that he was self-taught. Typically, Charlie was completely undaunted and not deterred in the least. He was on his own in Nice; none of the other British aviators were going to risk it, but Charlie had weighed up the pros and cons carefully. Perhaps he was being arrogant in his belief that his skill would match theirs. He was certainly not afraid to prove it.

Charlie had recently acquired the French-built Wright biplane, which was carefully packed up by Tommy and himself and loaded on to railroad transport to Nice. Charlie and Tommy were travelling by motor car, in Charlie's new Silver Ghost. They set off in high spirits but even before they reached Dover the police had detained them.

"Name?"

Charlie Rolls glared at the policeman who stood solidly, legs apart and braced, with pencil poised over notebook. The offending whistle dangled from his breast pocket.

"Offence?"

"Speeding on a public highway where there are restrictions in force!"

"Speeding? Oh Crikey! You call this speeding? My dear good fellow." Charlie paused and reached into the breast pocket of his thick tweed travelling coat and took out his cigarette case. He removed a cigarette, and slowly and deliberately took the new petrol cigarette lighter he had acquired in New York from the dashboard and lit it. He inhaled and blew the smoke towards the policeman. The policeman bristled and went scarlet in the face.

"Been lurking a long time in these bushes, have you?" drawled Charlie sarcastically. "I suppose these fines are crucial to the force, and presumably there is nothing else for you to do? Nothing useful? Possibly something in the public interest, like catch the odd thief?"

Tommy Smith cringed and sank lower in the passenger seat, sinking his head on to his chest, his hat masking a face overcome with embarrassment.

"Name?" asked the policeman doggedly, licking his pencil, his face set, mask-like, recognizing the driver.

"Smith!" replied Charlie, meeting the policeman's surprised look. "Let me spell that for you, S-M-I-T-H!" he spelt out slowly.

The policeman wrote out the ticket, not sure how to challenge Charlie's identity. He handed it over and Charlie took it silently, turning away from the policeman. He inhaled on his cigarette deeply and deliberately and slowly ground it out in the ashtray before driving away, ignoring the policeman and staring stonily ahead. Charlie could barely contain himself from accelerating at full throttle so that he could leave him splattered in mud. Speeding fines were a thorn in Charles Rolls' side. "Ridiculous," he muttered. "Forty miles per hour is not even fast. The Ministry of Transport is still living in the stone age. At least flying, they can't stop you for speeding."

Tommy Smith sighed resignedly and sat upright once again. That was the fifth speeding ticket he now had in his name. After a mile or so Charlie pulled in to the side of the road. "You might as well drive the rest of the way to Dover," he said to Tommy. "I need to think!"

Somehow, mused Charlie, public opinion has to be turned from the present conception that the pioneers were just a crowd of adventurers experimenting with machines that defied gravity. Somehow, the seriousness of the endeavours of the pioneers must be brought home to them. The very fact that the French, German and Russian governments had increased their investment in their own pioneers was a serious and threatening challenge and unless the British Government responded Britain would be vulnerable. Charlie believed the only way to engage the Government was through public opinion supported by the newspapers. Thus far, only Charlie Hands, employed by Lord Northcliffe for the *Daily Mail*, gave a real voice of support. Charlie believed that people like Lord

Haldane, Minister for War, Lord Montagu in the House of Lords and Lord Tweedsmuir, First Lord of the Admiralty were a real threat to national security. Charlie began to form an audacious plan. Much would depend on timing and the weather, but he felt he was ready and this international flying competition would be a good preliminary test.

The journey across the channel was uneventful. Tommy, as usual, took care of everything. His precious aeroplane travelled in advance and was to be unloaded from the cargo ship back onto a cargo train at Boulogne bound for Nice, and would be waiting for him on his arrival down there. Charlie and Tommy shared the driving.

Tommy realized from the long silences that Charlie was hatching a plan; he knew him so well. Usually conversation flowed, discussing the intricacies of mechanics and improvements that might be made and Charlie's plans to form his own aircraft manufacturing company. Tommy wished that Charlie had a regular lady friend now that Miss Vera Hedges Butler had married. What a waste. Tommy was glad he'd been able to marry who he liked. The rules of the aristocracy were nothing short of cruel. Kids sent away from their parents, had to marry within the strict confines of their class, couldn't show any emotion. Blimey, he was glad he hadn't been born into that class. Good-looking, decent bloke like Mr Rolls, had plenty of lady admirers but no one special now. He'd have made a good father; nippers always liked him. He was a good boss, always fair and caring, respected good craftsmen. Though he said it himself, he was a good mechanic, and there wasn't much T.O. Smith couldn't deal with.

Mr Rolls had a lot of respect for Mr Horace Short and all, as well as his brothers, especially young Mr Oswald. Shy bloke, Mr Oswald, but ever so sound, never tried to pull the wool or anything like that. Funny how Mr Rolls got on with so few of his own sort. It seemed that his best mates was Mr Brabazon and Mr Cecil Grace, both wild men, thought Tommy privately, which was curious when you thought how careful Mr Rolls tried to be on the whole. Still, everyone thought he was bonkers coming down here, but if you knew him like he did, then you'd know he wouldn't try anything unless he thought he'd got a reasonable chance. It was a brave gamble all the same; no one else was here to take the risks against these foreigners.

It wasn't the first time Tommy had been to Nice and the Riviera, of course, but it always gave him a special thrill when he got down there. He turned into the Promenade des Anglais and breathed in the warm seductive Mediterranean air. The tall palm trees lining the promenade swayed lazily in the soft wind, their dark green fronds a sharp contrast to the bright sapphire sea and white sand. It was a bit different to Leysdown's dark sea. Expensively dressed ladies and gentlemen glided leisurely along the promenade with all the time in the world on their hands. It was a different world down here. He drove slowly up to the front of the Negresco Hotel where Mr Rolls always stayed. The head doorman came forward hastily, immediately recognizing his guests, and clicked his fingers to summon footmen to help unload the luggage. Mr Rolls' reputation in France was greater than it was in England. Funny that, thought Tommy, how a lot of foreigners should think more of him than his own. He couldn't see a French Gendarme stopping Mr Rolls for speeding, much less booking him.

Tommy always expected to be put up in a local Bed and Breakfast, but Mr Rolls always insisted that he was accommodated in the same hotel as himself, so he was close by should he need him. That was another funny thing about the boss: as mean as mean he could be on some things and yet here he was in a posh hotel. As soon as he had settled in, Tommy drove straight off to the railway station depot to check up on the aeroplane. He took the precaution of taking his phrase book along with him.

Tommy had been to the depot on numerous occasions when they had shipped the Gordon Bennets down to compete in various races, so he knew his way around. He found the stationmaster and showed him the documentation. The stationmaster appeared flustered as he threw his arms about and gabbled so fast that Tommy thought he was having some sort of seizure. It began to dawn on Tommy that the aeroplane had not arrived. He pointed to his fob watch, trying to discover when it would be expected. The stationmaster once again threw his arms about like a windmill, and finally shrugged his shoulders.

"Je ne sais pas."

Tommy understood all too well that little phrase, and went cold despite the warm Nice climate.

"Où est?" he inquired, not caring about what accent to use.

"Je ne sais pas," the stationmaster repeated, shrugging his shoulders before waving his arms about and blasting another spurt of unintelligible French then turning abruptly and making off before Tommy could summon up another phrase. Tommy hurried after him, and for the next hour remained doggedly by his side, firing "Où est" and "Qu'est-ce que c'est?" at the stationmaster every ten minutes or so.

Meanwhile Charlie paced up and down outside the Negresco, waiting impatiently for Tommy. Patience finally ran out and he hailed a horse and trap to the station yard. When he arrived he could see an agitated Tommy shaking paperwork at a despairing stationmaster. Charles spoke fluent French, and when he heard the stationmaster's statement that his aeroplane was "perdue" he could barely contain his fury.

"What's up then?" said Tommy as Charlie let off a stream of invective.

"They have lost my aeroplane, and I have told the good stationmaster what I shall do if he doesn't find it within the hour!"

Twenty-four hours later there was still no sign of the aeroplane. There was nothing for it but to go to the flying fields and watch the other fliers compete. Tommy was deeply suspicious that they had been sabotaged. He didn't trust none of these foreigners, they was all smiles one minute, then, if you was the one what won they weren't quite so matey. Especially the Germans; there was no sense of humour there. At any rate, that's what happened with the motor car racing meetings. He didn't half admire Mr Rolls, cool as cucumber he was, watching everyone else and missing out himself. No telling what was going through his head, mind. It seemed the only person he took it out on was the head waiter at the Negresco. Sent his dinner back three times and in the end walked out. All he had was a few lettuce leaves. Ever so fussy about his food. In all the years he'd worked for him he had never seen him eat meat.

"I'm going back down the yard, Sir," said Tommy, "see what I can find out." Charlie nodded, his eyes firmly fixed on Blériot's Voisin biplane coming in to land.

"Well," thought Charlie, suddenly quite calm, "no sense in getting agitated. I need to watch and learn all I can from the ground and put it down to experience." Let Tommy do the worrying instead; he was obviously convinced they had been sabotaged. Charlie was sure that was not so, because everyone wanted to see each other's aeroplanes. Everyone wanted to see how skilled or otherwise their visitors were, and he had already picked up some very useful information. No, it was just the utter incompetence of the French railway system.

The stationmaster saw Tommy arrive, and to Tommy's relief he saw there was a beaming smile on his face as he came hurrying over to him.

"Ah voilà! It is found! It will be 'ere tomorrow!"

Two days out of the six had been lost. The initial excitement was replaced by a cool determination that he would succeed in every event over the next four days, and show the elite of France what he, a self-taught pilot, was made of. Tommy was proud of his boss; he made no fuss and handled his biplane in a businesslike way, completing trips from Nice to Antibes and back, and coming fourth against pilots who had had plenty of practice. Overall, even though he had missed two days of the event, he managed to come sixth, which was remarkable. There was nothing dashing about his flights. His sheer determination and courage helped to achieve his success.

When Charlie returned to Eastchurch the following week, Frank McClean held a lunch party at Mussel Manor in his honour, to which Horace and Oswald, Cecil, Brab and his new wife Hilda, Griffith and his old friend Frank Hedges Butler and Vera were invited. Charlie was secretly flattered, especially as Frank had ordered a special fish menu. Inevitably, there was no small talk. Horace asked probing questions about the engines and the other fliers, Oswald quietly took notes, while Charlie knew that they would minutely examine the information later. Competition was inevitably very fierce. Charlie looked around the table, studying each person in turn, and it came to him how truly fortunate he was that he could count on these people's loyalty and friendship. He smiled, and a familiar surge of ambition coursed through his body. He stood up. Everyone looked mildly surprised; Charlie Rolls did not often propose toasts.

"Madam! Gentlemen! To all aeronauts, but especially to those sitting around this table!"

He sat down abruptly, and for a minute there was silence, so uncharacteristic was the toast. Then Frank McClean stood and responded, and gradually, in turn, everyone proposed a toast, and there was a general noise and a laughter and feeling of warmth and well-being.

"What's next?" called out Brab, leaning back in his chair, grinning, with his hair flopping over one eye as he sipped vintage brandy from Frank McClean's cellar.

"I've decided I'm going to do a return flight across the Channel in April!" announced Charlie, his eyes glittering in the strange way that they did. "Furthermore, I shall tell Lord Northcliffe, who I'm sure, in his own inimitable way, will leak the news to the Government."

"Well done and good luck," the assembled company all exclaimed with excitement, especially Vera, who had not seen Charlie for some time; but naturally, as the co-founder of the Aero Club, no one questioned her presence at that exclusive gathering.

The news spread like wildfire. Charlie Hands haunted the Billet and the Castle, accosting villagers and Aero Club members alike for news of a definite date. Charlie Rolls and Tommy Smith shut themselves away in their workshop near Standford Hill; only Horace and Oswald were allowed entry. Charlie practised flying with Cecil whenever he felt the coast was clear. After work he would visit Rosewarne at Rosebank, join Horace there for a beer and stay on for high tea. He never grew tired of nursery food, eggs, cheese, fish and home baking; this was the diet he favoured. He often found himself eating on his own with the children after Horace had gone home and Rosewarne had returned to his building yard.

Once, just after he had made the decision to attempt the cross-channel flight and when his companions around the dining table were the Short children, Florence Cunningham and Dorothy, Charlie told them candidly that he was going to fly across the English Channel and back. They listened round-eyed with wonder and with a total belief that he would succeed.

"Will you take us up in your aeroplane one day soon?" asked Dorothy.

Charlie looked at all of them. You could not fob children off with "one day!" or "I'll think about it!" He considered the request seriously.

"No!" he said shortly. "Not for a long time yet, not until we are sure that the aeroplanes we fly are completely safe."

The children groaned.

"When will that be?" persisted Leonard.

Charlie shook his head, thinking of an honest reply. Dorothy studied

Charlie's face and looked deeply into his eyes and for some unaccountable reason began to cry. She excused herself abruptly and left the room. She ran down the corridor to the front door, let herself out and ran across the road, disappearing into the long grass in the field opposite, sobbing uncontrollably and not understanding why.

Leonard fired questions at Charlie, covering up for one of Dorothy's turns. Charlie was aware that there was something wrong, but he answered Leonard's questions as if nothing had happened. He sensed that Dorothy was not crying because he said he wouldn't take them up in his aeroplane for a long time.

By the beginning of May Charlie, felt he was ready, so he and Tommy took themselves off to Dover and camped in a field on the cliffs. Charlie Hands learnt of their impending departure over a pint in The Billet at Eastchurch a few days before they left. As a result, all the press were alerted, and reporters streamed into Dover. The only address Charlie had was "The tent near the aeroplane". Their journey was in vain. Reporters and spectators alike toiled up the cliffs above the town day after day, only to be told that the flight was postponed. Excuse after excuse was given for the delay but the fact was that the little team of Charlie, Tommy and a dear friend, Louis de Silva, were confronted by thick cloud or drifting fog, day after day. Charlie was not going to risk it, for he couldn't afford to damage his aeroplane. The cost was more than he could bear to lose, but most of all he was determined to succeed. Ambition had never made him reckless – he knew that that was a road to disaster – and fear was something Charlie never considered. One locked that away in a cupboard, under the stairs or in the attic.

Spectators came from miles around and waited as time and again Charlie was not satisfied that the conditions were right. Once he actually began a short flight, but no sooner had he taken off than he realized that something was wrong, and a crash was narrowly missed when a bystander panicked and ran across his path. The drifting fog finally got to the reporters and spectators alike, and just like the fog they drifted away back to London and back to their homes all over the country, disgruntled and disillusioned once again with the whole flying circus. The Honourable Charles Rolls had been consistently rude and bitingly sarcastic to anyone who was brave enough to ask him when he thought he might be able to make this historic flight. It was little wonder that they gave up waiting for the right moment.

Only the intrepid Charlie Hands stuck it out. He knew that Charles Rolls would never give up his attempt at the return crossing, but he would never go until he was sure conditions were right. Charlie Hands had learnt a lot from the times he spent at Leysdown and Eastchurch. He set up his own tent. He was not going to miss this flight if it was the last thing he did. Charlie Hands was there for a month, and it was the most uncomfortable month of his life. Right ho! If they could do it, then so could he. He lost weight, he got a streaming cold, he knew he was smelling and he vowed that he would never camp again, ever, in the whole of his life whatever the ground-breaking story might be.

The morning of 2nd June Charlie Rolls awoke with a feeling of confidence that that day he was going to accomplish what several eminent scientists said was impossible, and the Government had dismissed as a vain hope.

Charlie Hands woke at dawn and could hear the revving of an engine across the field. Cursing and swearing, struggling out of his sleeping bag still fully dressed, he wrestled with his shoes, grabbed his notebook and ran towards Charlie Rolls' aeroplane, coughing and spluttering, Tommy observed Charlie Hands as he stumbled across the field. Give the bloke his due, he was the only one who had stuck it out, not only the shocking weather, but his boss's biting sarcasm too. He poured a cup of coffee from the jerry can into a tin mug and held it out to Charlie.

"Thank you, Tommy," said Charlie Hands gratefully. Tommy nodded and walked back briskly to Charlie Rolls, who was now settled, in the Short Wright aeroplane with the Voisin engine which Charlie had taken with him to France. It was hoisted onto its starting rail, the engine revving, but once again the mist closed in. The day dragged on and it seemed that once again the weather would thwart the attempt, but by mid afternoon the weather conditions seemed to ease and then without warning Charlie decided to make the flight at 6.30pm guided by both instinct and a favourable weather report. As the mist swirled about the edges of the cliff, suddenly the engine was revved more strongly and Tommy and Louis released the starting rail. Smooth as silk the little machine took off and disappeared over the cliffs into the mist. Sure enough, the mist cleared halfway over the Channel, and his flight to Sangatte and back went according to plan even though his vision on the return journey was masked by fog before he came to the Kent coast and he was forced to use the sun as a guide. Otherwise the flight was reported as being "uneventful". It took forty minutes each way and ninety minutes in total, averaging a speed of 30 mph. He was more or less averaging the same time as Blériot's flight in 1909 and Jacques de Lesseps in May 1910, but neither had flown as many air miles as Charlie.

Charlie Hands' frustration and tiredness of the past month disappeared in the realization that Charlie Rolls was making history, and that he was the only reporter to witness it. He ran down the slopes into the town to the post office and alerted everyone, including Horace, of the news. Within an hour he was back up at the top of the cliff, as was most of the town of Dover, nearby Folkestone and all the surrounding villages and countryside hamlets. The news had spread fast, the minutes ticked by and people straining to see an approaching aeroplane sometimes mistook birds in the distance, and a groan of disappointment would go up when they realized that they were mistaken. Then, exactly ninety minutes from the time Charlie Rolls had taken off, far out to sea, a small dot appeared, and with a distant drone like a bee, it grew and materialized into an aeroplane. Charlie flew a circuit around Dover Castle before coming in to land. Overnight he became a national hero.

That evening, Charlie and Tommy had hoisted the aeroplane on to a trailer which they attached to the Silver Ghost, and made their way from Dover to Hounslow. It was very late by now, and they were very tired and Charlie was in no mood to be flagged down by the police force, who queried the legality of towing their aeroplane behind the Silver Ghost. Charlie's response was icy, telling the unfortunate policeman that when he could produce a copy of the Highway Code that forbade aeroplanes on trailers, then he would do them the honour of complying with the law. He bade them good evening and drove off fuming.

Their reception as they entered the village that evening after they had unloaded the aeroplane at Hounslow was overwhelming. The villagers, alerted of their approach, lined the streets and cheered, waved and whistled. Charlie was smiling and exhilarated, but anyone could see that he was almost embarrassed by the attention. He retreated into the vicarage while Tommy drove the Silver Ghost down to their Standford Hill workshop. Dulcie Dickson gave him a warm embrace and all the children, Leonard, Ruby and Francis, Dorothy and Florence, seeing him go there, came racing in from the street where they had been part of the reception crowd. The children jumped up and down with excitement.

"You did it! You did it!" they cried. Charlie just did not know what to say.

Later on, when the crowds had dispersed and the reporters realized they were to get nothing more out of Charlie, they melted away back to London. He left his temporary sanctuary and joined Tommy in the workshop. They packed up the various tools they needed, and early the next morning they made for Hounslow for their next exhibition.

Charlie's rise to National Hero status was quite overwhelming. Messages poured in to South Lodges, where, installed in his new office, Miss Caswell valiantly coped with the deluge of fan mail. The Royal Aero Club presented him with a gold medal, an award that had only been given to Wilbur and Orville Wright, Louis Blériot, Henri Farman and Hubert Latham, thus becoming the first Englishman to receive this award. King George V sent a congratulatory telegram, Madame Tussaud's commissioned a model in wax, the press forgot their vendetta with "the noble scorcher" and the *Daily Mail* wrote: "This double crossing of the Channel is an encouraging indication that this country is waking up to the real importance of airmanship."

The *Daily Chronicle* wrote of "a daring feat brilliantly performed" and its leading article said "Mr Rolls has wiped out a reproach which attached to us as a nation for our backwardness in the new art of aviation."

The *Pall Mall Gazette* portrayed a cartoon with John Bull in which he says "Well done my boy, and thanks. You've given me a lift which I rather badly needed."

The Motor expressed the view that "Britannia Rolls the Waves" and the London Vegetarian Society expressed their delight that the hero of the day was a vegetarian. And just to add to all these plaudits a statue was erected of Charlie on the seafront at Dover looking out towards France.

CHAPTER TWENTY

Bournemouth – a great career is lost

Charlie had no plans to slow down, and although he still retained his main workshop at Eastchurch, he had also purchased a shed at Hounslow which he found convenient for South Lodges. The Wright needed a good overhaul before his next appearance at Dunstall Park, and as usual he had Rolls-Royce work to catch up on, even though he was taking a lower profile in the business to allow him to concentrate on the advancement of the aeroplane, and the training of new fliers and mechanics.

The Aero Club knew that this was their opportunity to maximize Charles's return cross-channel flight. So, hailed as a national hero, Charlie volunteered his services to train pilots to fly and Cecil did likewise. It was a particularly magnanimous offer on Charlie's part given the huge demands he already had with his responsibilities to his Rolls-Royce Company, and his own most recent exploration into aeroplane design. Unbelievably the offer was met with an indifferent response from both the Government and the War Office.

Disappointed by this negative reaction to his offer, Charlie then set his sights on Bournemouth's centenary celebrations, which were to last for two weeks at the beginning of July, and promised to attract dignitaries from a very wide spectrum. Patrons headed by royalty were expected to attend. The Lord Mayor of London added his name to the endless list of VIPs, but the most significant of all, from the pioneers' point of view, was the Secretary of State for War, Lord Haldane and his Under Secretary, Lord Lucas. Once again, Charlie and Cecil offered their services to the Admiralty to train pilots free of charge and, to their amazement, this time they jumped at their offer and it was agreed that recruits could form up at Eastchurch after the Bournemouth Exhibition.

Bournemouth had gone to town on the arrangements for the centenary celebrations, and had done everything in its power to make sure that it was going to be a memorable fortnight. The *Daily Telegraph* wrote enthusiastically on 6th July:

> For the better part of a fortnight tomorrow Bournemouth will positively be rioting in revelries and festivities. Where it has been possible to improvise a temple of gaiety and mirth, wherever greensward or sheltered glade seemed likely to serve the purpose of the celebrations, there have the sleepless organizers staked out their claims.

Masked balls, parades of decorated cars, march pasts of Boy Scouts and cadets of the Royal Navy, Royal Marines and Merchant Navy were all planned, and a new aerodrome was created in the suburb of Southbourne. The programme of events at Southbourne attracted nineteen entrants, including pilots from France, Germany and Belgium.

Charlie was as usual planning ahead while at Bournemouth. He had discussed his future plans with Tommy that day, telling him that following the Bournemouth celebrations he intended to take some time to act as an instructor to Army pilots at Laffan's Plain, and then he also wanted to visit America to compete in The Garden City Meeting, where he hoped to meet up with Wilbur and Orville Wright again. He also discussed his new ambition with a young aviator called Keith-Davies and revealed his plan to form his own aviation company, offering the young aviator the opportunity of "getting in on the ground floor".

Unfortunately, Charlie's overhaul of his Wright Flyer for the Bournemouth meeting had not gone altogether as planned. He had decided to install on his aeroplane (in the words of C.H. Gibbs-Smith) "an additional rear outrigger built to surround the existing rudder outrigger and to carry – at first – a fixed tail-plane to give a measure of inherent longitudinal control inevitable with the front elevator". This device did not work, so a second elevator had to be manufactured, and this did not arrive until 7th July and was not fitted for another three days, so that Charlie had no time to test it out before the meeting started. Charlie was obliged to travel by train to Bournemouth as the evening before he had collected yet another speeding ticket in Hyde Park.

The first day's programme was fairly uneventful. Charlie did not manage to lay claim to any prizes, and Cecil managed to crash his Short Wright. The second day's programme held the challenge of the quick starting and landing contests which Charlie had set his heart on winning. A one-hundred-yard circle had been marked on the ground with a large white patch marked in the centre, which the aviators had to aim for. Unfortunately the circle had been placed in a most impractical position, approximately sixty yards from the grandstand, and approximately sixty yards from the judges' box, and the marks were not at all distinct, so much so that Charlie asked Tommy to stand near the area so that he could see where he was meant to go. On top of this the wind had risen, and was officially recorded as anything from 17 to 25 mph. Other pilots had misgivings, and Edmond Audemars was so concerned he tried to dissuade Charlie from making any attempt until the afternoon when the wind may have decreased. The time then was 11am.

Charlie had already made one attempt at the landing competition, but with the wind factor hampering the final approach there were three choices: the first was to approach downwind and risk overshooting and colliding with the judges' box; the second was a cross-wind approach; and the third was to meet the wind head on and come in to land with the nose facing the grandstand. At 12.30 Charlie shook hands with Sir Thomas Lipton from his Wright Flyer and taxied out into the arena.

In the words of *The Motor*:

> In little more than two minutes, he had curved over nearly half the aerodrome inside the far pylons, and was making the necessary turn behind the line of the grandstands. This was the only way open to him, since he had decided to face the wind in the completion of his pre-alighting manoeuvre. He turned gracefully and well, apparently under perfect control, at a height of some one hundred feet, and brought his aeroplane round over the R.A.C. end of the stand, and above the heads of those in the enclosure until it was making directly for the point at which he aimed. Standing vertically below him as he passed above the barriers which separated the enclosure from the

> flying ground proper, I instinctively realized that he was keeping too high for a comfortable finishing angle. His altitude above the barrier may have been seventy feet. I believe that Rolls had refrained from making an earlier considerable alteration in his angle of descent while he still was over the enclosure and between the lines of the grandstand and the barrier, wholly out of splendid consideration for the spectators beneath him. As soon as he was over the aerodrome proper, he started to descend sharply.
>
> There was a sickening snap. Some part or parts of the tail-plane had given way just inside the line of the barrier – witnesses who had profile views say that a failure of some kind had already occurred when the biplane was over the enclosure, but there was no evidence of it in the flight when viewed from above. An involuntary spasm of the nerves went through one's system, and then there was a thud. Rolls was down under our very eyes, it seemed but a few paces away.

Within minutes his friends had warded off curious bystanders and photographers and formed a tight circle around Charlie. There was not a mark on him, and it took some time to persuade his friends that he was actually dead. He had suffered concussion to the brain and in fact had died instantly. It was the most appalling shock, and it is probably true to say that his closest friends, Brab, Cecil, Tommy and Frank Hedges Butler, never got over it completely. And nor did Vera, Frank's daughter. No one will quite know or understand why she married Iltid Nicholl when she did, but the family are in no doubt that Charlie was indeed the love of her life and, at one time at least, this love was reciprocated. It is fair to say that the popular description of Charlie's character by Lord Montagu as a cold individual, incapable of making lasting relationships, is totally inaccurate; one only has to read accounts of his character from his true friends such as John Moore-Brabazon, Cecil Grace, the Shorts, Francis McClean and the intensely loyal Tommy Smith.

Lord and Lady Llangattock were on their way from London to Poole to board the *Santa Maria* and were taken off the train. Charlie was taken to a nearby mortuary and although flying was halted for the rest of the day the other festivities continued, the people unaware of this dreadful tragedy.

There have been many theories as to the exact cause of the crash, but Tommy Smith and Keith-Davies agreed that not only was he flying too high to make a safe descent, but in lowering the elevator and going into such a sharp dive, the outrigger and the second rear elevator were unable to cope.

All those who had celebrated his historic return flight over the Channel a month before now deeply mourned his death. *The Autocar* summarized Charlie as "Earnest, persistent, courageous and modest. Charles Rolls was the stuff of which the best Englishmen are formed, and his country can ill afford to lose such in the prime of their lives."

The Motor: "That so careful, so unassuming, and so quiet an experimenter as Rolls should have been called upon to pay this price seems very hard."

The Daily Mail: ". . . one who assisted in removing the fetters on mechanical invention. Rolls died a martyr to knowledge: his memory and example will be an incentive to continued effort in an art that crowned him with victory, and claimed him as her sacrifice."

The Society of Motor Manufacturers and Traders: "The whole industry will always stand indebted to him as one who rendered the very greatest assistance to the movement at a period when so many difficulties and so much prejudice had to be overcome."

A memorial service was held for Charlie in London, to which all his friends, members of the motor trade, the press, members of the Government, royalty and friends from France and all over Europe flocked to pay tribute. Charlie Rolls was an outstanding pioneer and visionary whose contribution to society has never been fully appreciated, and now he is virtually forgotten.

His friends felt the loss dreadfully. John Moore-Brabazon announced that he would never fly again, Cecil Grace refused to believe he was dead, Vera collapsed and never really recovered, and Lord Llangattock was affected more than anyone realized. He knew that there were few who really understood his extraordinary son, certainly not his wife or many of his contemporaries; he loved him deeply and understood him better than anyone. He died only three years after Charlie was killed.

Mr Eustace Miles (fellow advocate of food reform): "I never heard anyone say anything against him except that he had too much to do."

Of all Charlie Rolls' obituaries, that of Charlie Hands, who followed Charlie Rolls' career from the outset, is the one that tells us quite plainly what a great man Charlie Rolls was and lays bare his personality and his remarkable prophetic gift in addition to an extraordinary pioneering scientific talent.

Type of Young England

by Charlie Hands

Poor Charlie Rolls! We who knew him realize that England has lost a great deal by his death. For he was a shining example of that spirit of activity and effort which is the answer of the young of the race to the apathy and "don't care attitude" that a few years ago seemed, after a generation of easy prosperity, to have possession of the mind of the nation.

I knew him in his motoring days as something a good deal more than a mere motoring enthusiast! It was not motoring that fascinated him. It was engineering.

I saw him learning to fly in his first Wright machine. It was not that he had lost interest in motoring and had caught the new infection of flying. He was still the same devotee of engineering and mechanical science. I did not know him when he was a cyclist, but I know that he was the same man in those early days. It was the mystery and fascination of the machine that even then caught his imagination. It was the hungry desire to do something in the world under the condition of accomplishment prevailing that led him to follow the development of the mechanical science that holds the greatest possibilities of accomplishment.

He had no occasion to do things. He might from his young days have looked forward to a careless life of easy enjoyment of a world in which nothing needs to be done. He might, if his inclination had led him that way, have found an outlet for his energies in sport. He might have hunted and shot in the winter and played golf and cricket in the summer.

But there was for him no satisfaction of achievement in any pastime. He had a burning passion to do something that counted. Nothing more untrue or unkind could be said of him than to describe his interests in life as being those of a "sportsman". When he had done something with his working day he might have passed a leisure hour in the enjoyment of some sport, but he was no mere sportsman. It was achievement and enjoyment that he sought in life.

Once I accompanied him on a motor run from London to Paris. His interest in the journey from London to Southampton and from Le Havre to Paris was not in the towns we visited, or the scenery we passed through or the people we encountered. It was the engine of his big racing car that interested him all the way.

On the road we had serious trouble with the engine. One of the four cylinders of the big racing Panhard with which he had cut the kilometre record in England a few days before went wrong somewhere along the road between Le Havre and Rouen. He managed to get the limping motor car into a small village.

The mishap brought him not disappointment but one of the delights of his life. For it was necessary to pull the big engine to pieces in order to get at the seat of the trouble within the cylinder. He had no tools for such an undertaking, but in the village he found a wonderful French mechanic in a workshop with a lathe. The two of them put on their overalls and got to work upon the complicated machine. They took it all to pieces, found out the trouble, duplicated the damaged part, and built up the engine again in workman-like style. He had started out to break the French kilometre record. That, under the circumstances, was impossible, but Mr Rolls derived a delight from the successful repair of the complicated engine far greater than any sensation of pleasure he might have derived from cutting the speed record.

Ever afterwards he kept up close friendship with that village mechanic who on several occasions visited him in London, and whom he went many times to see when occasion carried him to France. The love and knowledge of engineering made a bond of sympathy that brought the young English aristocrat and the French workman into a close and enduring friendship.

He took up flying as he had taken up motoring, not for the enjoyment of the sensation of flight but because he was eager for the new opportunities of the engineering achievement which airmanship offered. In motoring he had done more than break records. He had evolved the wonderful and beautiful Rolls-Royce six-cylinder engine, as perfectly finished a piece of mechanism as a chronometer.

It happened that I was with him when he was learning to fly and I believe I saw him make his very first short flight at Shellbeach of a quarter of a mile. It was a science, not merely sport, that marked those early experiments. With patience and determination he applied himself to mastering the difficulties of flight. He had practised with a gliding machine before he obtained possession of the first Wright aeroplane that was seen in England. He had been to France to see one of the Wright brothers fly. Down upon the marshes of the Isle of Sheppey, the scientific thoroughness with which he had applied himself to work he had taken in hand compelled the admiration and respect of everyone who saw him.

If he had wished merely the adventurous joy of flying, he might have made extended flights long before he did. He restricted himself to short hops of a couple of hundred yards or so because it was not a sporting sensation he was aiming at but a complete mastering of the new science. He knew every part, every wooden strut and wire stay of his machine, had calculated the strains and stresses that would be developed upon every bit of material. He was an engineering enthusiast, not a flying man.

When he was a boy at school and at Cambridge he used to be called "Dirty" Rolls by other boys who saw no fun in getting their hands blackened and their cloths saturated with oil. But he did not mind that. He applied himself to his academic studies, took his degree with distinction, disdained no branch of knowledge – but always, and first of all he was an engineer, because he wanted to do something in the world and it was engineering that offered him the fullest opportunity.

In airmanship he would have succeeded in developing something as notable as the Rolls-Royce Motor. All his flying experience was directed to that end.

"It is no good to be something," I once heard him say. "What is wanted nowadays is a man who can do something and do it well."

All his life he tried to do something and do it well. Young as he was, he was one of the first Englishmen to ride a motor bicycle and drive a motor car. He was a great reader, familiar with the modern literary developments in Germany and France as well as in England.

He was abreast of progress in other branches of science as well as his own. He lived his life in earnest. I saw him yesterday lying dead among the wreckage of a great career.

CHAPTER TWENTY-ONE

Prisoners of hope

Hilda Brabazon knelt before her husband, and looked into his ravaged face, afraid for him, herself and their unborn child. Every now and again she wiped away the tears trailing down his cheeks before they landed damply on his already wet collar. It was four days since Charlie's accident and still Brab wept. He shook involuntarily, reliving the moment over and over again when he had heard the sickening snap and watched his friend plunge into the enclosure before the eyes of all the spectators in front of the stand. There was not a mark on his body, as he had been flung free of the debris. He looked perfectly at peace, with a half smile on his face. Tommy was the first to reach him, and when he and Cecil got to him seconds later, Tommy's stricken face told them he was gone. From then on Brab, Cecil and Tommy were forced to shield Charlie from the public and prying newspaper men, and they were cool and calm in refusing to allow anyone near him. Brab and Cecil, still hoping he wasn't really dead, just in a coma, carried him into a small tent near the judges' box. The band in an adjoining field were playing one of Charlie's favourite tunes, "The Dollar Princess". When one doctor pronounced him dead, Cecil dismissed him and called for another doctor.

"Hey! He's not dead!" said Cecil firmly.

Lord Llangattock brought his son's body home to Monmouth. His death affected him more than anyone realized. He knew there were few who understood his extraordinary son, certainly not his wife or many of his contemporaries. Most members of the Aero Club, along with Horace and Oswald, travelled down to Monmouth to attend the funeral. Some of the villagers attended, among whom were Dulcie Dickson and her husband, and Charles Rosewarne. The Honourable

Charles Rolls was buried in the family plot and Lord Llangattock commissioned a memorial in Agincourt Square, Monmouth, shortly after the funeral took place. It was a long way for anyone to come and pay their respects, and the family plot itself was isolated, accessible via a long, winding lane, out of sight of the public. Who would lay flowers on Charlie's grave in ten, twenty, fifty, a hundred years' time?

Horace, Oswald and Charles Rosewarne returned from the funeral by train to Eastchurch. It had been a long and tiring journey, with several changes of train en route. They sat opposite each other in silence, each lost in their own melancholy thoughts. Rosewarne had offered to drive to Monmouth but in the end they had decided that it would be better to go by train. Oswald returned to Mussel Manor, where he had lived on his own for some time. It was pleasant enough in the summer, but he determined to move into Eastchurch village in the Autumn.

The influx of fliers into Eastchurch meant that Oswald had more work than he could handle, so in some ways the solitude suited him. He could work uninterrupted on his paperwork in the evenings, but at times he felt lonely. He was too shy to make the short journey to Eastchurch and visit The Castle or The Billet. Eustace continued to live in London and worked out of Battersea, keeping his distance from Horace. Tonight Oswald was glad of the peace and solitude of Mussel Manor. He was going to miss the enigmatic Charles Rolls. Oswald believed his aloofness was a façade to disguise his passion, and his powerful visionary talent where there was no room for sentiment. He had been driven by a force so strong that it shut out everything except the vision ahead. Oswald understood completely, for he himself knew how compulsive it was for him to invent and create.

When Horace opened the front door of Parsonage Farm he was greeted by the familiar smell of home cooking. Marie Claire came out of the kitchen to greet him, complaining that he had not told her when he was returning, then asking innocently in French how his trip had gone, completely forgetting he had been to the funeral. Horace exploded unreasonably, all the pent-up emotions of the previous week boiling to the surface. He looked at his wife dressed all in black, a stocky figure with her hair greying; all her original attractiveness, it seemed to Horace, had disappeared. She was no longer the woman he had married. Marie Claire took a step backwards, afraid of the look on Horace's face.

"You take no interest in anything but the kitchen," he shouted furiously.

Marie Claire retaliated. "What else do you want me to do? Tell me? I cook for you, I clean for you, I see to our children. What more do you want?"

Leonard and Francis shared a room and they were woken by their father's voice and came running down the stairs, alarmed.

"Get back to bed!" Horace roared, turning in their direction. His head was throbbing.

"I want more than a housekeeper," he said more quietly, trying to calm himself.

Early the next morning he went to see Charles Rosewarne, who was having breakfast with some of his workmen around the ebony table where Charles Rolls had sat so often in the early evening with the children.

"Sit down, Horace! Cup of tea? Eggs? Bacon? Black pudding?" inquired Rosewarne cheerfully.

The men finished their breakfast, got up from the table, and left. Horace accepted a mug of tea. Rosewarne waited for Horace to speak. He knew from what Kate said that all was not well with the Short household. Horace said nothing. He didn't know how to start, nor indeed, now that he was here, what Rosewarne could do in any case. Charles Rosewarne looked at Horace, realizing that he had come to talk about some difficult problem which he now felt unable to broach. He had a brainwave to break the ice.

"You're a swimmer, ain't you?" he said.

"Yes?" replied Horace curiously, taken by surprise.

"Tell you what you can do next Saturday, take you out of yourself and that, enter the annual race across the Swale; mind you, I've won it now three years on the trot!"

Horace felt himself smiling, the first smile he'd had for nearly a week. Physical exercise, just what he needed. A bit of training in the Sheerness baths, and he'd go and cheer himself up and see the lovely Edith he'd met recently at the Britannia Hotel.

"Well done, Rosewarne! Just what I need; and let me tell you, you won't win it this year!"

Horace didn't have time to get into Sheerness that day. When he got to his workshop Oswald was already there, and also Frank McClean was waiting for him. Everything was happening so fast that Frank was concerned they would be overwhelmed with visitors who were merely curious. On top of which, new aeronauts were moving into the already overstretched community. What concerned Frank most, however, was the pressure the Short brothers were shortly going to be under to develop their Short biplanes, and he didn't want them to break under the strain. He knew that both Horace and Oswald had taken Charles Rolls' death very badly, because even though it was not their engine, they were horrified that they might have been held responsible, especially Oswald. Frank felt it was his duty to impress on Oswald that he should never ever blame himself should one of his aeroplanes be involved in a fatality. It would be a worse tragedy if Oswald ceased his work because the strain of responsibility of other people's lives overwhelmed him.

Horace and Oswald listened to Frank's concerns and his insistence that they must never consider consequences which might turn out to be ill fated. That was not the way to progress. They must put the tragedy of Charles Rolls' death behind them and move on; he would have done the same, for his motivation was always to move forward whatever the cost.

"Kid!" Oswald no longer minded Horace calling him "Kid", as he now took it to be a term of endearment.

"You must get Eustace down here. I need him," said Horace.

Oswald had never thought he would hear those words from Horace. Oswald nodded, aware, nevertheless, that Eustace would not like being torn away from his balloons in Battersea. However, he had a good manager who could take charge. Eustace would see sense in the end and Oswald could see the urgency and necessity of being alerted to increased activity in every area. Part of him felt excited and part of him felt overwhelmed. Could they meet the challenge; could they deliver?

Horace had to wait until the end of the week before he managed a trip into Sheerness. He made straight for the Britannia Hotel. It was six o'clock in the evening. Plenty of time for a swim in the public baths. They didn't close until nine o'clock at night in the summer. Edith was busy at reception, her head bowed, working out bills or menus or some such thing. Horace walked quietly up to the desk. It was good to see her; she looked as voluptuous as ever.

"Hello, my beauty!" he grinned. Edith's face lit up with pure pleasure. She threw open the little hatchway, let herself out of her booth, and flung her arms around Horace.

"Where have you been, handsome?"

"Flying about," replied Horace, and laughed at his own joke.

Edith stood back and looked at him. "Well, you look done in I must say. Come on, let's go and have a nice glass of claret in the snug and you can tell me all about it, ducky!"

Horace never got his swimming practice that evening. Instead he went to the baths first thing in the morning and got back to Parsonage Farm for breakfast. If Marie Claire was suspicious she didn't show any signs. It was the school holidays, and Leonard and Francis were still in bed.

Either the swim or Edith had done him good. Horace didn't know which, but he felt a new energy he hadn't felt for a long time. Reinvigorated, he returned to his new engineering design for aeroplane engines with increased energy. Oswald glanced over to Horace's design bench and was pleased to see Horace looking more relaxed than he had of late.

"Horace!" called Oswald over the hum of hammering, saws and machinery. "Eustace said he will come down here next week."

"Excellent," replied Horace and began humming to himself.

Whether it was morbid curiosity or genuine interest was difficult to tell, but people descended on Eastchurch in their hundreds to see for themselves the flying fields and the pilots experimenting. The villagers closed their doors, surprisingly hard hit by the Rolls tragedy. Some felt guilty that they had originally been so sceptical. An exhibition had been arranged, long before Bournemouth, to take place in the autumn, not on the same scale, but nevertheless there were a good many entries not just from Eastchurch but from Brooklands, Hounslow, France and Germany. The villagers had been looking forward to it but now no one wanted to talk about it, and it seemed too soon to get the bunting out again.

Horace and Oswald were pleasantly surprised when Cecil came to call on them one morning. He stood shyly in the entrance. He was always rather overawed by Horace; he took his cap off and held it with two hands in front of himself. He was the same age as Oswald, and, like Oswald, looked younger than his twenty-seven years.

"I need to get in the air again," said Cecil quietly after they had exchanged small talk regarding the funeral and the weather.

"Good!" exclaimed Horace. "So you should, well done! We'll check over your aeroplane and tell Frank you are here, he will be delighted! Now what about Brab? Will we see him soon?"

Cecil sighed and shook his head "I'm afraid not. I've tried to talk to him but he is adamant, and he says he will never fly again!"

"What?" said Horace and Oswald at the same time, shocked.

"I know; maybe it's partly Hilda who doesn't want him to fly. She is expecting a baby soon, and I believe that that is part of the reason. Anyway, he has told me that that is his firm decision and I do not believe he will change his mind."

There was an awkward silence. Horace looked down at his drawings and with his pencil absent-mindedly drew a small pig in the corner of the sketch board, remembering Brab's defiance at a report which had asserted that pigs might fly but not men. So Brab had put a pig in a basket and flown the first round mile with the pig beside him to earn £1,000.

It was hard for Cecil returning to the old flying ground, knowing his friend would never be around again. However, just like riding a bike, you had to get up and carry on. After all, he did not have the same personal responsibilities as Brab. It was so essential to keep going and show everyone the way forward. Dammit, he was going to do his bit, yes sir! The Government were coming round, he could see that. Why, this young fellow Churchill was pushing and pushing them to expand and develop all the time. He sure hoped his offer to train pilots was being taken up. The country needed all the pilots they could handle. It was obvious that Germany was spending more and more on aviation, and what for? It surely wasn't for fun. No, sir!

Cecil found he turned to Amy Cunningham more and more. He had increasingly come to rely on her wise advice and counselling. He found that he could talk to her about any subject under the sun with an ease which he was unable to find with anyone else, friends or family. She had a razor sharp brain and an astonishing grasp of current affairs for a woman of the era. She believed in so many of the things he believed in. She had secretly joined the suffragette movement. Why, her husband would surely divorce her if he found out, and he wasn't sure how her friend Kate Rosewarne would react. Of course it would be great if her husband would divorce her, but that was highly unlikely, unfortunately. It was the only forbidden subject they had between them. Amy had helped him so much over the last two weeks to recover from the shock, sorrow and loss of his friend. She was so tiny, yet so strong. She was ten years older than him and yet she still looked like a young girl. Their age difference was completely irrelevant so far as Cecil was concerned. And it wasn't just her incredible horsemanship or her pure genius as a musician that attracted him – it was Amy herself. There was nothing they could do right now except be grateful for the time they managed to steal together. Maybe one day . . . ? When he looked into Amy's eyes he knew that everything he felt was being reciprocated.

Cecil kept his horse in Frank Ingleton's stable at Stone Pitts. He was going there now to saddle up his horse and ride over to meet Amy on the cliffs out towards Minster. It was a beautiful day and Amy would probably bring a picnic, which they would eat on the beach before splashing about in the sea. John Cunningham still spent Monday to Friday in London. Cecil took his leave of Horace and Oswald.

"Before you go," called Horace "if you are not doing anything next Saturday, come and support Charles Rosewarne and me at the annual gala swim across the Swale."

"My word," exclaimed Cecil laughing, half out of the shed. "I wouldn't miss that for the world, sir!"

Eustace arrived a few days later, and within minutes he and Horace had disagreed about their divisions of work and how they were going to proceed.

"Look!" said Oswald. "We're after the same goals, so let's just work our priorities out. We've still got the usual cash flow problems but pretty soon we know the Government will come knocking at our door, so let's be as ready as we possibly can for them."

Horace and Eustace eventually managed to settle down and work side by side. Horace worked on intricate engineering drawings, aiming to replace existing engines in their aeroplanes, while Eustace worked on the much larger designs of an airship, which was a tentative enquiry for an order from the Admiralty. Horace bowed to Eustace's superior knowledge of balloons, which was relevant for the construction of any such aircraft.

Oswald went back to the Leysdown sheds to work on other sections. It was convenient for Mussel Manor, and that week Captain Dunne had returned to continue his work on his D5. The tailless V-shaped biplane, with sharply swept-back wings and a rear-mounted engine that drove two pusher propellers, was now much more refined and Oswald had undertaken an order for the two wings. Oswald found him a wonderful companion. Apart from his obvious talent as a designer, like Oswald he loved the countryside. He was a knowledgeable ornithologist and taught Oswald a great deal. After work they were able to explore the hedgerows and cliffs and marshes, although Oswald was never able to share Dunne's enthusiasm for shooting and, like Charlie Rolls, he had become a vegetarian. Both Dunne and Oswald, like the other pioneers, also planned to move to Eastchurch, as soon as it was practical.

Frank McClean had sold the Shorts a three-acre site for £25, and they had commissioned Charles Rosewarne to build them flying sheds which they worked from for the next four years.

Eustace accepted Horace's offer of his spare room at Parsonage Farm for three nights a week, returning home to Battersea for the remainder of the week. Eustace felt sorry for Marie Claire. She seemed very pleased to have him to stay, but her inability to speak English was a big drawback. The children translated for him. At first Eustace was too tired to pick up the straining relationships within the family. But then he did think it was curious that Horace spent so many nights away from home.

It soon struck Eustace that Marie Claire was frightened of Horace, or rather Horace's temper, and she did not know how to deal with it. Fortunately, she had one good friend in the village, Kate Rosewarne, who was virtually a neighbour and spoke reasonable French. The children had struck up good solid friendships between themselves but Leonard, Ruby and Francis were certainly aware of the strain on the relationship between their parents. There was more than just the pressure of work at stake.

As soon as the news was out that Charles Rosewarne was racing against Horace Short, the fierce-looking one of the three brothers, excitement gripped the whole island. Charles Rosewarne was their champion, and a challenge from an outsider had never happened before. Fred Waterman, the Eastchurch butcher, ran a book, and bets were being placed with high stakes. News spread through the fliers' and mechanics' camps as well, and men called in at Fred's shop before the race to place their bets.

Charles Rosewarne practised in the sea at Leysdown early every morning and every evening. Dorothy and Florence accompanied him, like trainers.

"Come on, Papa . . . come on, Mr Rosewarne!" they yelled from the beach. "FASTER!"

Horace went into Sheerness and wasn't so sure that his training was quite up to scratch, especially as he was always keen to spend part of the time with Edith, and there was no doubt that she was a bit of a distraction.

"You will come and watch, won't you, sweetheart?"

"Course I will. I wouldn't miss my hero's triumph for the world. I shall go with friends so no one will be suspicious. Don't you worry!"

One morning, a few days before the race, Leonard, Ruby and Francis cornered Horace at Parsonage Farm.

"Dad!" said Leonard seriously. "You've got to go and practise in the sea like Mr Rosewarne. He goes twice a day with Dorothy and Florence down to Leysdown. People are taking bets on who's going to win, Dad!"

"My word!" said Horace, genuinely astonished, and decided he'd go down to the Swale that very afternoon. Never mind the sea, he needed to know what the Swale was like to swim in. He would go to see Edie later.

Eustace was not impressed at being left for swimming practice. He was going to miss the swimming gala in any case, as he was playing football for his local team. Eustace considered football a proper sport, like boxing. Swimming was not a proper sport in his eyes, and he wasn't taking time out for extra training. It was ridiculous to take a swimming race so seriously, especially when there was so much to do. But Eustace was in the minority; everyone else was excited, and really pleased to have a distraction from the recent tragedy.

There was a real festive atmosphere as the crowd swarmed down to Kingsferry and crowded on to the bridge. The swimmers assembled below. Kate and Marie Claire had declined to attend; Amy and Florence went together, and Cecil came with his brother Percy, keeping their distance even though John Cunningham had stayed at Shurland. There was no point in attracting any gossip. Oswald saw Horace's boys and struggled through the throng to join them. Dorothy scrambled her way in and out of the people, gradually making her way to the centre of the bridge for a prime viewing position, then squeezed her way to the front. Fred placed himself at the finishing line while Bert Chesson, Rosewarne's joiner, and his wife were the other side of the bridge, with others of Rosewarne's employees, near the starting line. Horace spotted Edith dressed in bright turquoise with a matching parasol. Even from where he was, below the bridge, he could hear her laughter. He looked up and caught her eye, grinned and winked. She gave a tiny wave with the fingers of her right hand. Her elaborate reticule dangled from her wrist; the sequins caught the sunlight and sparkled sharply, attracting Leonard's attention. He caught his father's exchange with the gaudy-looking woman with the glittering reticule. He turned away and yelled at his father, waving a bright yellow handkerchief through the iron railings of the bridge. Horace was still looking at the woman and didn't see Leonard's frantic waves or hear his cries of "Good luck!" Leonard wasn't the only one who saw the exchange between Horace and the woman dressed like a peacock. Oswald wondered who on earth the woman was, and what on earth was Horace's connection with her. He looked around fearfully for Marie Claire.

Dorothy was screaming at her father and waving a red handkerchief. "Papa, Papa – good luck!" Charlie heard and blew a kiss, grinning from ear to ear, and waved back to her. The crowd starting cheering as the swimmers lined up. Twenty competitors in all waded across the dried-up mud banks of the Swale into the murky-looking river. In fact the water was surprisingly clean, and fairly clear away from the banks. The starter blew his whistle and off they plunged, Charlie and Horace kept a sedate and rhythmical pace in the middle, about a yard or so apart, eying each other as they lifted their heads on every fourth stroke. They were in no hurry, for there was a long way to go, but it was clear that they were both equally strong and powerful swimmers. On the turn the pace quickened and suddenly they had left the rest of the field behind.

Reaching the centre of the bridge, Horace was ahead by half a body length. The crowd groaned, thinking that their champion was going to lose. "CHARLIE!" yelled the crowd as loudly as their lungs allowed them. "HORRY!" screamed Edith, beside herself, attracting hostile stares from all around her. Horace went ahead by a whole body length. The crowd groaned again. Horace lifted his head to shout at Charles triumphantly. It was a mistake, because Horace thought the race was won, and in that ill-judged moment he lost his rhythm. Charles, quite unperturbed, took a deep, deep breath, lowered his head and accelerated, drew level with Horace and reached the finishing rope in front by a second. The race was Charles's but Charles declared it a draw, for they had both broken the existing record.

The children rushed down to greet their fathers, their pride bursting. They had outstripped all the other swimmers and apparently broken a record in the process. Oswald saw Edith looking at Horace, and as inexperienced as Oswald was in these matters, even he recognized that the woman knew Horace better than as a mere acquaintance. At that moment Horace searched the crowd to find her. He smiled up at her, she blew a kiss and turned and walked back across the bridge escorted by two other men towards a waiting pony and trap.

Horace could not relax, and was totally committed to new developments. His head reverberated with new ideas and designs now that he had finished all the Short Wright biplanes. He was exhilarated at the way progress was being made, and buoyed up by Frank McClean's enthusiasm and that of the new arrivals. Grudgingly, he respected Eustace's and Oswald's work, aware that he was hard on them. But they had to push forward, they could not afford to slow down now, and every mechanical detail had to be meticulously checked and double checked. He was probably being unfair on them, because he knew perfectly well that they did so. It was, after all, their excellence in workmanship in their balloon factory that had got them this far, Horace reasoned. He must be fairer, he told himself.

When Horace travelled the short distance from Parsonage Farm to the Stone Pitts farmland every day he could see the activity intensify week by week. Sheds were being extended, and everyone was learning to fly. Even Frank McClean was now an able pilot.

Horace and Eustace had managed a fairly amicable truce since they had started working together, until one day Eustace announced that he felt he should learn to fly. Horace flew into a rage at the very idea.

"What if you have an accident?" he roared. "Who is going to teach you? What are you going to learn on?"

Eustace had no intention of being bullied by Horace; he had already made arrangements to take flying lessons with Cecil and told Horace so; and furthermore he was not going to cancel them just for Horace. Their voices rose and the technicians and pioneers nearby smiled and rolled their eyes; they had already witnessed some of the brothers' previous heated exchanges.

Cecil Grace, Chilean born and a graduate of Columbia, was the son of American citizens. His father, John Grace, was a New York banker, and when the family moved to London Cecil then took a place at Oxford. When his father died he made his home with his mother in London. It was easy to understand why he had become such a good friend of Charlie Rolls: for both men had a strong pioneering spirit, they shared the same fearless sense of adventure and sense of humour (as shown in the flight from Leysdown to Eastchurch and the message "The Cage is Empty, the Bird has Flown"). Cecil had been awarded the fourth flying certificate from the Royal Aero Club. He obviously felt enormous affection for his adopted country, and became a naturalized British citizen in October 1910. He pursued countless challenges, not only in England but on the continent, but in December 1910 he had his sights firmly set on the Baron de Forest prize of £4,000, for the longest flight from England into continental Europe, and was sure that he had every chance of succeeding. Cecil was delighted with Horace's modified designs of the S27 pusher biplane.

Cecil departed on 22nd December from Swingate Downs, informing Amy that he would be back that evening. It turned out that there were strong winds and great swirls of fog which hampered the flight. Eventually he made it to Calais, but there he decided that he would postpone the prize flight for another day and return to Eastchurch via Dover instead. He sent a telegram to say that he had made it to

Calais and that he was going to return via Dover and would be leaving Calais at around two o'clock in the afternoon. In fact he left at ten past two. The flight should have been no longer than forty minutes but at 3.30pm there was no sign of him, although an aeroplane had been sighted by a coastguard at Ramsgate at 3.00pm heading north about six miles out to sea.

For days it was hoped that Cecil had been able to land somewhere, but when a pilot's goggles and cap were washed up at Mariakerke in Belgium and were identified as belonging to Cecil, all hope was lost of his still being alive.

Cecil Grace was an enormously popular member of the Royal Aero Club and hugely respected for his achievements in the extraordinarily short time he had as a pioneer. He was posthumously awarded the Gold Medal of the Royal Aero Club "for his achievements as a pilot and competitor". He was also commemorated by the Aero Club de France when his name was inscribed on a monument they erected celebrating the earliest cross-channel flights.

Charlie Rolls and Cecil Grace were true "*Prisoners of Hope*". A stained glass window in All Saints Church, Eastchurch, unveiled in July 1912, is dedicated to them in memory of their pioneering gifts to the nation.

The memorial window unveiled in July 1912 in Eastchurch parish church to Charles Rolls and Cecil Grace, Sheppey aviation pioneers and the first British pilots since the advent of manned flight to be killed in flying accidents

CHAPTER TWENTY-TWO

Forgery pays off

Although Eustace and Horace had come to a reasonable working relationship for some months, the pressures of their inventiveness were taking their toll. After a particularly stormy morning Eustace left Horace to rant and rave on his own and went off to find Oswald at Leysdown. It would be good to have a talk with him as there were a lot of things he wanted to discuss, not just work. There was something very wrong with family life at Parsonage Farm, and he was worried more for Marie Claire's sake than his volatile brother's. Oswald was surprised to see Eustace; it was the middle of the morning and Eustace rarely took a break at that time.

"Shall we go and have a coffee at Mussel Manor?" inquired Oswald, sensitive enough to realize that whatever had made Eustace take time away from the work shed, it was quite serious. Of the three of them Eustace was the most stalwart, turning down help and refusing advice when he could easily have asked for it. He nodded agreement. Oswald rolled down his shirt sleeves and took his coat from the hook on the back of the door of his small office. They walked over to Mussel Manor in silence, Eustace trying to work out which was the more pressing subject, Marie Claire or the order for the seventeen gas bags for the amphibious air ship. Or perhaps the fact that he could no longer stay at Eastchurch, as there was no room to build anything which needed space, like air bags. They sat in two armchairs in the window of Mussel Manor looking out towards the sea.

"Good of you to spare the time to call, Eustace," said Oswald encouragingly.

Eustace took several mouthfuls of coffee, then put his cup down on a small table beside his chair.

"I needed to talk to you about several things as a matter of fact. Have you seen Marie Claire recently?" he asked.

"No," said Oswald slowly, vividly recalling the woman on the Kingsferry bridge.

"I do not think that all is well with Marie Claire and Horace. They do not seem to be, I don't know", he paused, "getting along any more, and the boys are very unhappy. She never seems to leave the house and seems permanently afraid of Horace, while Horace appears to be permanently irritated with her . . ." Eustace trailed off and looked at Oswald for help.

Rightly or wrongly, Oswald decided not to tell Eustace about the woman he had seen at the Swale Swimming Gala.

"Why don't you persuade Marie Claire to take the boys and go and see mother in London for a long weekend? Perhaps, maybe that would help. She has always been fond of the children and often complains she doesn't see enough of them. There are lots of things for them to do at their age in London: the National Geographical Museum, they would enjoy that and the Zoo and Madame Tussaud's; that is an amusing place to visit, I hear."

"What a good idea," replied Eustace. "They have been stuck here all summer and in fact ever since they arrived; it's high time they had a break. Then perhaps when Marie Claire returns, Horace will have missed her and the children, and everything will return to normal," said Eustace optimistically.

"And . . .?" Oswald knew Eustace well enough to know there was something else on his mind.

"I cannot stay at Eastchurch, Oswald. No, it's not entirely Horace," he said, smiling ruefully. "There are two important contracts from the Admiralty which need space that not even Battersea can handle."

Quietly the two brothers discussed the recent orders for seventeen air bags for the amphibious air bus, and a very large observation balloon commissioned by Griffith Brewer for the Admiralty. They knew they had to handle these orders on their own. Horace would not be interested, and nor of course did he have the right experience in that field. Oswald firmly believed that gold beater's skin was essential for rigid airships, but Eustace told him that the Admiralty was unwilling to pay for it, and they were thus obliged to use the cotton and rubber fabric used for domestic balloons.

"Typical," said Oswald, shaking his head in mock disbelief. "But you are right, Eustace, we have to find an alternative manufacturing base. I believe we should go to Rochester: good access to water and I know there are some excellent new buildings along the waterfront to rent. We should make enquiries as soon as possible."

Eustace instantly agreed and they made arrangements to go there. He felt better for his chat with Oswald. He might be the youngest brother, but he was the one with the clearest head, and more and more Eustace bowed to his quiet reasoning and advice.

In the spring of 1911 Wilbur and Orville Wright made what was to be their final visit to England before the First World War. They visited Brooklands but had come primarily to visit the Shorts to see for themselves the progress they were making at Eastchurch, and to ask if they would manufacture their aeroplanes for them. Horace declined the offer, because by then he was confident in his own ability to design his own aeroplanes, and was keen to continue developing his own ideas, even though it was a most prestigious offer. They accepted his decision with good grace, but knew they had lost the services of the best manufacturer in Europe. Horace was charming but firm in his refusal. Oswald understood his brother's reasons immediately but Eustace disagreed with the decision, although he said nothing at the time.

Captain Dunne, Tom Sopwith, Richard Fairey and Geoffrey de Havilland had all come to meet Wilbur and Orville Wright. Of course at that time they were virtually unknown, and little did Wilbur and Orville know that in a few years' time they would all become famous names in aviation. Both Wilbur and Orville were touched at the fuss made of them, and amazed at the activity and intensity of the work surrounding them. Captain Dunne demonstrated his own aircraft, and amazed the Wrights with his original design, naturally unaware that this was actually what aircraft would look like in the not so distant future.

Oswald and Eustace showed the Wrights the innovative ideas and draft plans for their seaplane, and brought them to Battersea to their balloon factory, which ironically was as busy as ever with commissions from a wide variety of customers, including the Admiralty. They also manufactured and sold a variety of equipment such as gas bags, automatic valves and differential pressure gauges, all designed by Oswald. The Wrights certainly had their eyes opened to the diversity of designs manufactured by the Shorts. Their friendship was re-established once again, and Wilbur and Orville gave warm reciprocal invitations to Eustace and Oswald to visit Ohio. Sadly, during the visit, they could not help noticing how frail Wilbur appeared to be.

It seemed that all the good intentions of training army or naval officers to fly had been put on a back burner since the deaths of Charlie Rolls and Cecil Grace, who had both volunteered their services free of charge. The Royal Navy appeared to refuse to acknowledge the importance of their involvement with aeroplanes, which was baffling. The Admiralty actually wrote: "... we regret there are no funds available at the moment to train any of our officers in the art of flying ..."

Horace and Frank replied swiftly, offering to train four of their best officers and adding that Frank McClean would supply his own aeroplane to do so. Once again they pointed out Germany's enormous budget of £1.5 million on aviation. Frank got in touch with Charlie Hands at the *Daily Mail*, and explained that they had had no real positive response from the Admiralty so far as training officers to fly was concerned and they had had to write and offer free training to try to encourage the Navy to send them some. It would be most helpful, Frank concluded, if the general public were made aware of this situation.

Subsequently Frank was not sure whether it was his letter, or Charlie's splendid article, or pressure from their junior minister Winston Churchill, that had persuaded the Navy, at long last, to volunteer to discuss recruiting officers for flying instruction. Frank received a short cursory letter from the Admiralty, on which the signature was completely illegible and looked very much as though a clerk had been given the job on the instructions of his superior, and that the superior could not even be bothered to sign the letter himself. The letter merely said, "the Admiralty accepted Mr McClean's and Mr Horace Short's offer to train officers to fly, and would send a Lt Charles Rumney Samson to set out the terms".

The impertinence that they were setting out their terms, as if they, the Admiralty, were doing them some sort of favour, was typical. Never mind, it was a start. They only hoped that Lt Charles Rumney Samson was going to be suitably appreciative, and was not going to be some puffed-up arrogant officer of the old school.

A few days later, around mid morning, there was a rather timid knock on Frank and Horace's office door. "Come in!" roared Horace; the door opened to reveal the entrance of a diminutive and diffident young man dressed in naval uniform apologetically explaining why he was half an hour late due to the late running of the train from Victoria Station. Frank and Horace relaxed. However, it was short-lived; when Charles Samson told them the terms under which the Admiralty had allowed only four officers to train, they were outraged. The officers were given three months' leave of absence but must forego any prospect of ever commanding a ship. They had to be single and be prepared to pay out of their own pockets for any damage they might incur to the aeroplane. However, despite these stringent terms, two hundred officers applied but only four were selected, including himself, Lt Arthur Longmore, Lt Louis Gerrard and Lt Reginald Gregory. Samson also told Frank and Horace that they really owed this concession to a Captain Murray Sueter, who was attached to the Admiralty, for persuading them to accept their generous offer. Captain Sueter worked extremely hard, badgered one of the Lords of the Admiralty, and in the end, deeply frustrated with their Lordships, forged something himself, worded suitably so that the powers that be would not question it. Horace and Frank exchanged silent acknowledgement that Captain Sueter was a valued ally.

Charles Samson had one particular favour to ask them from Captain Sueter but would understand if this could not be accepted: "He would very much appreciate it if he could also bring six army officers to Eastchurch that week. They would have already taken leave, and would train alongside the naval four. Captain Sueter believes that it is a matter of absolute urgency in light of the activities in Germany at the moment."

Horace and Frank readily agreed, and Frank told Lt Samson that Mr Short would be giving ground instruction and a Mr George Cockburn would give flying lessons. In the meantime they would find more aeroplanes, and he, Frank McClean, would also conduct flying instruction. He suggested that Lt Samson find digs in the village and said he would look forward to seeing him the next morning at five o'clock. When Samson took his leave and saluted them, Frank and Horace roared with laughter and told him it was not necessary to salute them.

Charles Samson appeared to be quite overcome in the presence of these two men. He was deeply aware of their remarkable reputations: the great Horace Short, the first manufacturer of aeroplanes in the world, and Frank McClean, one of the very first pioneers and the main benefactor, without whom none of this could have been achieved; both men willing to train members of the Armed Forces at their own expense, without any real support from this backward, intransigent Government, content to put at risk the country's very security. He had enormous respect for them, as did Captain Sueter.

It was a deeply depressing situation for Horace and Frank, wondering when the Government, the Admiralty and the War Office were ever going to back them. It felt as though Frank McClean was funding England's defence budget single-handedly. Horace made up his mind that it was time he invited his old boss, Sir Charles Parsons, to Eastchurch to show him exactly what was being achieved there without any support. He was sure that Sir Charles Parsons would use what influence he could to draw attention to the dangerous attitude of the Government.

It was curious that six army officers were coming to Eastchurch to be trained alongside the naval officers, as in the spring of 1911 an Air Battalion had been set up to take over all the aeronautical activities of the Army, covering work on airships, kites and balloons and as a result the Army Balloon Factory at Farnborough became the Army Aircraft Factory. At last, Prime Minister Asquith had asked Lord Haldane to examine the whole aviation situation, and he came up with a plan to amalgamate all the existing small flying organizations into one called The Flying Corps, with five sections: a Naval Wing, a Military Wing, a Reserve, a Central Flying School and the Army Aircraft Factory; and in 1912 the Royal Flying Corps was created by Royal Warrant. The young Geoffrey de Havilland was their aircraft designer, and their engineer from the Daimler Company was one Frederick Michael Green. They produced their own aircraft as the most suitable for development, and it was at this stage that the Navy's approach did not coincide with that of the Army. This was profoundly important so far as the Shorts were concerned, because the Royal Naval Air Service was eventually formed from the original Naval Wing of the Royal Flying Corps, ensuring that the Navy kept their independence from the Army. This meant that they were able to establish their own purchasing policies and deal directly with whichever manufacturers they considered produced the right aircraft or seaplane to the best standard and specifications for their use, such as Shorts, Vickers, Blackburn, Sopwith and later Fairey rather than the Royal Aircraft Factory.

It was Captain Murray Sueter whom the Shorts had to thank for his great support and influence and uncanny perception in preserving and protecting the smaller manufacturers from being swallowed up by the state aircraft factory. He fought long and hard for his views to be taken seriously by the Sea Lords. In his own words he commented that the Sea Lords,

> in their great wisdom, would not be guided by their airmen, but preferred to take the views of an elderly ex-First Sea Lord on the Committee of Imperial Defence, who could see no value in airships. When this Admiral was opposing airships and crabbing their development for the Navy, I am perfectly certain he had never seen one.

Sueter continues,

> I could get little money for air development before the War. One Sea Lord sent for me and kindly said: "Seventy-five thousand pounds have to come off the Naval Estimates this year. So of course, it must be taken off "Air" – and it was!

Murray Sueter was a magician in spreading his appallingly small budget to allow the Shorts to manufacture their seaplanes. He records:

> The money available was small, but I do think, when considering the whole matter many years after, that our firms gave us really wonderful machines for the very little financial encouragement we were able to give them.

It must be said that it was as a result of Horace's and particularly Frank's enormous generosity in reciprocating the Navy's faith in them that there were gains on both sides. Frank McClean generously gave the use of the airfield at Eastchurch free, and he also paid for the construction of two more Short biplanes. George Cockburn, one of the first Englishmen who learnt to fly, offered his instructing services free of charge.

The increasing numbers of curious sightseers visiting Eastchurch and begging for a flight were becoming a nuisance, interrupting work and disrupting routines, so that restrictions on visitors had to be imposed to avoid countless repetitive tours. Nevertheless, they were keen to show the public what they were doing, and certainly did not want to be secretive in any way, as they wanted to encourage as many as possible to accept this new form of transport in any way they could.

Of course it was difficult to vet every visitor, but one day, as Horace strolled over to a small queue, he noticed that the pilot was looking particularly agitated and that a mechanic was attempting to restrain a young lady from attempting to board the small aircraft. The pilot quickly took Horace aside and explained that when she attempted to board he noticed that she had a long knife tucked into her boot, and, given the fact that she also had a mad glint in her eye, he definitely did not want to take anyone for a flight carrying such a lethal-looking knife. Horace told the pilot to leave it to him; he would deal with her. He took her arm and whispered something in her ear and then, without any protest, propelled her across the airfield towards some of the sheds. The display continued until the last visitor had departed, when Horace appeared from a different direction with the young lady, strolled slowly back to the anxious pilot and handed her over to him, pulling his jacket quickly open so that he could see the offending knife safely retrieved and tucked inside Horace's jacket. She climbed into the aircraft quite meekly. The pilot took her for a circuit around the field and landed without mishap. As she walked away, a man and a woman came to meet her and escorted her to a waiting car. Sometime later, Horace learnt that she had escaped from a lunatic asylum. No harm was done, and no one ever learnt how Horace retrieved the knife!

Early Short Aeroplanes:
Short No. 1 – tailless pusher biplane. Short No. 2 – Machine in which J.T.C. Moore-Brabazon won the 'Daily Mail' £1,000. Short S27 – first Normal Box-Kite. Short S38 – the aeroplane on which Naval Pilots received their flying instruction. (Courtesy of Medway Archives)

CHAPTER TWENTY-THREE

Frank McClean makes a daring flight through the middle of Tower Bridge

In May 1912 the sad news reached them that Wilbur Wright had passed away. He had died of typhoid, his delicate constitution unable to stand the potency of the disease, despite the best medical treatment available. He was forty-five years old, and would be remembered for his gentle character, his keen intellect, his modesty, his inventiveness and above all his incredible contribution to the world. A memorial service was held in Washington later that year, when members of the Royal Aero Club, the French Aero Club and pioneers from all around Europe attended to show their deep respect.

Work continued unabated in Eastchurch. Horace, Eustace and Oswald were only human, and for all their contribution to aeronautics, their temperaments were far from compatible. Eustace and Oswald got on well enough, but the three of them together were explosive. Everyone knew that you disturbed Horace Short at your peril, and his quick temper and his lightning tongue left the bravest mechanic quaking. Equally, his keen sense of humour and his love of practical jokes was disconcerting. No matter what mood Horace happened to be in, he nevertheless was held in high respect by everyone who came in contact with him, and mechanics were willing to wait all day for five minutes of Horace's time. The most endearing quality Horace possessed was that he was totally unaware of the awe in which he was held. When the date for Wilbur's memorial in Washington was announced he did not hesitate to persuade Eustace and Oswald to go, despite the overwhelming workload.

Although Horace had written to Sir Charles Parsons inviting him to visit Eastchurch, he was amazed and flattered when his invitation was promptly accepted. Sir Charles spent several hours in Horace's work shed examining detailed drawings of the original designs of the Short Wright biplane, and up to his most recent designs of his seaplane, not yet fully in operation. Sir Charles saw for himself Horace's complete dedication to his work, his inventiveness and his sheer genius. He was also made keenly aware of the respect in which he was held by other fliers, mechanics and the recently arrived officers from the Navy and the Army. In addition, there was obvious affection and concern displayed by many of the fliers, who worried he was over-stretching himself mentally and physically.

When Sir Charles returned to Northumberland, he sat down and wrote to Mr Asquith strongly recommending Horace Short for a knighthood. He decided not to tell Horace for the time being, certain that Horace would not take it seriously and could even sabotage his own chances with his modesty, making light of his own very real achievements.

Inevitably Horace's marriage came under an intolerable strain. Marie Claire's absence for a week when she took the children to London to stay with her mother-in-law had had no effect on Horace's temper. In fact, she doubted if she and the children had been missed at all. It was not just the missed suppers, the unexplained absences, but she knew, on top of everything, that Horace had a mistress. It was so obvious. But she was after all a Frenchwoman and she would never complain about a thing like that. No, it was the neglect that they suffered and the terrible rages which frightened her and terrified Ruby and Francis while Leonard withdrew into himself. Children could not cope with such rages and nor could she.

Horace himself wished he could be more tolerant, as it was depressing having such an unhappy family life. In the event, he was blind to the fact that Marie Claire had had enough. One morning she methodically packed up everything she needed for herself and the children to tide them over until she had found a suitable place to live, and telephoned her mother-in-law and asked her for accommodation for herself and the children until she had made other arrangements. At the time the children believed this was a temporary move and Leonard told Dorothy that they would be back soon. This was not to be the case. Fate played a part in their ultimate move: Leonard had won a place at Rochester Grammar School and that is where Marie Claire decided she would move to. She knew that Oswald and Eustace were trying to negotiate a move there for their business and it seemed that this would give the family the support they needed.

Of course when Horace finally returned to Parsonage Farm, late, on the day his family had moved out, he was deeply shocked and sharp pains of remorse made his head throb. He knew in his heart he could not put matters right. Now, more than ever before, his work had to come first, for the very lives of generations to come were going to be at stake if he did not give everything he had to the future of the aircraft industry. His poor family was the sacrifice he had to make.

He dreaded the condemnation he knew he would receive from his mother, Eustace and "The Kid". Perhaps "The Kid" would understand in years to come. Even now it seemed that Oswald, not yet thirty years old, was dedicating his life to the development of air travel for the progress of mankind. Horace wondered if Oswald had already recognized the grave responsibility he had undertaken, and realized subconsciously that there was no room for a family of his own. What a dreadful loss to the female gender, thought Horace wryly, a young man who was kind, talented, and blessed with movie star looks. Eustace was a different kettle of fish and a dark horse, thought Horace. He tried to rehearse what he was going to say to Eustace. Perhaps Eustace would never speak to him again. He drifted off to sleep.

Horace made the decision to stay on in Parsonage Farm so that when and if the children came to see him there would always be somewhere for them to stay; it would be their second home, or so he hoped. Later that week he called on Edith. Edith put the world to rights for Horace, made him laugh and made no demands. Horace rented a small cottage on Old Rides Farm, between Eastchurch and Leysdown, where the two of them spent their precious spare time.

Despite Horace's volatile private life, his work and inventiveness continued to blossom. The Shorts owed a great deal to the support and faith in them shown by members of the Aero Club, and particularly by Frank McClean; when cash flow was looking grim Frank would invariably order another aeroplane! They were also grateful to Captain Murray Sueter, who never wavered in his loyalty and belief that only the Shorts could produce seaplanes to his demanding expectation of excellence.

From around 1912 the Shorts' main activities were concentrated on the production of seaplanes. Oswald describes in his own words their main operations in Eastchurch at that time:

> We built the first twin-engine machines that were ever built. This was a two-seat Tractor Biplane with two Gnome engines in tandem, one in the nose driving two side propellers by means of chains, and the rear engine driving a pusher propeller. This machine was, therefore, the first machine with three propellers. Mr McClean ordered this machine of Horace's design, which was known as The Triple Twin. We built also a twin Gnome engine Monoplane Seaplane with twin floats and land chassis, really an amphibian. The floats were airbags built of canvas lined with rubber balloon fabric.

Oswald also explains:

> It was natural that, with naval men under instruction at our works, our thoughts should turn to the subject of making machines to alight on and get off water. I think we may claim to be the pioneers of the tractor-type seaplane with twin floats, a type of machine which has held its own to this day [1925], and I believe still has a great future before it. We were certainly run close, for by then Commander Schwann [also a noted naval aviator], all unknown to us, was engaged in fitting twin floats to an Avro tractor aeroplane at Vickers, Barrow-in-Furness. His first attempt, however, was not successful and the machine was smashed. It is said that this attempt was made towards the end of 1911. Our first seaplane [the S-41, later the H-10, then simply 10] was being constructed then, and flew in April, 1912 (piloted by Lt. Samson). Its first flight was at Grain Island, a mere hop, immediately previous to its being sent with the Fleet to Weymouth, where there was a Fleet Review (in May) by the King. There had been no time to give it a proper test. Lt Samson flew the machine at Weymouth on May 12, going out to meet the King's Yacht

> as it arrived: he flew for a considerable time and made flights during the period of the review.
>
> This machine had a 100hp Gnome engine, and carried about six hours' worth of fuel, a pilot and a passenger. The floats were made with a wooden girder structure braced with wire (like a fuselage); they had a step about midway, and two air holes to ventilate the step. The bottoms were of quarter-inch three-ply wood, and the sides and top were covered with rubbered balloon fabric. The wing-tip floats were inflated air bags made of rubbered balloon fabric attached to a skid. The tail float was of similar type. The writer [Oswald] was responsible for the design of these floats, and ever afterwards the fitting of floats fell to my lot.
>
> Of course, I had the advice of my elder brother in this work when I required it. The weight of the floats was the great problem then, and we had to save every ounce, hence the fabric sides and the inflated bags for twin-tip floats and tail float.

In June 1912 the first world record was set by Lt Samson, who had now been promoted to Captain, flying the first seaplane 195 miles in 194 minutes from coast to coast.

Other methods of landing on water had been explored with a Shorts 50hp Gnome Pusher Biplane, with wheels and skids, and fitted for a water landing with two bags of rubber fabric inflated with air attached along one side to a skid which in turn was attached to the aeroplane skids by a short length of steel wire. The total weight of each bag was 8 lbs and still lighter bags were fitted on the tail booms to prevent overbalancing on the water.

Oswald recalls:

> Many of those who knew this experiment was to be made believed that these bags would certainly burst on striking the water. One "expert" (they were already around in those days) called them "soap bubbles". Lt Longmore began to take an interest, however, when he saw them being tested one day suspended on a pole and filled with water. They held over 700 lbs of water, that being their displacement. Just as the bag was full of water a fastening slipped and the bag fell four feet or so to the ground, but did not burst.

> Longmore was impressed, but before the flight was carried out he and I carried out a further test by dropping the bag with 200 lbs weight attached to the nose, a distance of 30 feet or so into water from the pier at Queenborough. The bag withstanding this test, Longmore was satisfied, and the flight and landing on water was carried out quite successfully.
>
> This was the forerunner of machines fitted with light air bags for landing on water, much used during the late war for ship aeroplanes and used quite recently as a safety device for commercial aeroplanes crossing stretches of water. One of our machines similarly fitted [S-38] was sent in HMS Hibernia (a second class cruiser) together with our first seaplane to the Weymouth Naval Review in May 1912, and the former machine [S-38] flew from the deck of the Hibernia, and landed on the ground outside Weymouth.

This was the first flight from a moving ship's deck.

Horace's and Oswald's biggest problem at that time was the fact that they had to transport the seaplanes to Queenborough, which was the nearest suitable water. They had to be completely dismantled because the lanes to Queenborough were narrow and had high hedges either side of them. But Horace soon overcame the problem and designed folding wings, so that after that the aeroplane rested comfortably on the water-launching trolley for the journey.

By the summer of 1912 there was still a frustration among many of the fliers that most of the Government, and indeed the general population, were still not taking aviation seriously, and the advances they had made were still not being given the attention they deserved. It was typical that when Frank McClean flew from Eastbourne to Eastchurch on 17th June an action for damages was brought against him, alleging that "the defendant did so unskilfully and negligently fly over Spratts Farm that a valuable mare was frightened". However, undeterred, Frank, sacrificing his own career in astronomy to instruct naval officers to fly, also undertook to instruct Territorial Officers as well. However, the War Office issued a directive that he should abandon this as "they did not consider it advisable for Territorials to learn to fly", which left everyone utterly dismayed at such shortsightedness.

At this point Frank, someone who was not usually depressed by events, was at a low ebb, and he missed Charlie Rolls; Frank knew that Charlie would have known what to do and he would have drawn the public's attention to the matter in his own flamboyant way.

Fresh from this disheartening command from the War Office, Frank abandoned his instruction to Lt Longmore for the day, and walked briskly and angrily through the small airfield and climbed Standford Hill to look out towards the marshes and the Swale, misty in the summer's heat. He could almost hear Charlie Rolls' curses as once again he tried his first flights from the top of that hill. The Swale wound around the marshes like the thick roots of a tree, as the tendrils, like the dykes, spread out into secret, treacherous channels sheltering the marsh birds, harbouring water voles and protecting the territory from human intrusion. It was a magical place and the background of extraordinary history, past and present, reflected Frank wryly. He followed the haunting cry of a wild egret, swooping and diving over the marshes in search of food. The sound of an aeroplane drew his gaze away towards Elmley, and for all the world they appeared similar in the distance. He could just make out the iron structure of the Kingsferry bridge and watched the egret disappear from his view in that direction to fly underneath it before soaring up into the clouds and out of sight. The aeroplane also disappeared from view heading in the direction of Minster, no doubt completing a test circuit.

Frank did not know if it was the egret, the aeroplane or Charlie Rolls' ghost which inspired him, but he walked briskly down the hill and made for Horace's office. Horace listened to Frank's plan, barely able to contain his laughter, while Frank finished explaining the audacious feat he was proposing.

It so happened that the *Daily Mail* had advertised a visit by a famous French aviator, a Lt Conneau, who adopted the alias "Monsieur Beaumont". His reputation was well known on both sides of the Channel, and he always managed to attract large crowds. His visit was scheduled for 10th August. Charlie Hands spent much of his time at Eastchurch, but that day he was in Fleet Street busily typing up a piece for a deadline he was not sure he could meet, when he was called into the sub-editor's office to take a call on the telephone.

"Hello, old boy," the familiar voice of Frank McClean hailed him. "I've got something for you. I tell you what it is, I've arranged to meet that frightfully famous French aviator 'Monsieur Beaumont' . . ." – Charlie had the odd feeling that there was a touch of leg-pulling in the tone of Frank's voice, but perhaps he was reading something into his tone which wasn't really there at all – ". . . on London Bridge in a week's time at midday. I would be much obliged if you could arrange a spot of advertising in advance. What do you think?"

"It sounds great, Mr McClean, yes . . . I'm sure we could arrange that all right."

"Excellent," said Frank enthusiastically. "We need a good crowd there; it should be quite an occasion."

"I imagine it will be, sir!"

"If I were you I'd try to get one of your boys a pass to the Houses of Parliament and get on to the terrace with a cracking good photographer."

"Right! So you would like myself on London Bridge, with a cracking good photographer, midday, a week from now, and another reporter on the terrace of the Houses of Parliament with another bloody good photographer?"

"That's it, old boy, in a nutshell!"

"Right you are!" Charlie was intrigued; to hell with the deadline. He owed a lot to Frank McClean. He'd get going on a piece right away. A telephone call secured his photographers but it took all afternoon to secure the pass to the Houses of Parliament. Parliament was in recess and many of the members were absent. Finally, after drawing a complete blank, Charlie made the excuse that the *Daily Mail* wished to interview a War Office official. He was given the pass and a minor official, whose vanity was flattered that a newspaper reporter should wish to interview him, agreed to meet the reporter to discuss present policies at 12.00 noon on 12th August, exactly seven days from now.

Charlie had a busy week writing spurious articles in the *Mail* about the mysterious forthcoming visit of the famous French aviator. Eventually his editor questioned these numerous articles, but Charlie was able to reassure him that, whatever was going to take place, it would be spectacular and only the *Mail* had the scoop. Meanwhile, back at Eastchurch, the aviators were put in the picture and instructed to spread the word to their friends, and as much as they could to the general public, to congregate on London Bridge or Westminster Bridge. At 11.55am precisely Charlie was poised. He observed with satisfaction the crowds around London Bridge and presumed there would be as many around Westminster, which in fact there were, as well as along the Embankment and around Lambeth Bridge. The official from the War Office was, of course, late but the reporter was there on time, aware that the interview was just a screen for something else. A few parliamentarians wandered in and out on to the terrace; they had spotted the crowds and wondered what was in store.

It was a wonderful day, acknowledged Frank. He had been smiling ever since he had got up that morning. Eastchurch was deserted, of course, as everyone had gone to London. Two mechanics remained to see him off. They checked and re-checked the Short Biplane S-33. The delicate open fretwork in a box construction supporting the double wings housed the tiny cockpit where the pilots sat behind the engine, whose propeller was as large as the overall height of the machine. The tail of the aeroplane was also an open fretwork in a box-like construction. Underneath the cockpit and the wing there was what can only be described as a pair of skis, with which the plane could land on the water. It was certainly a unique design, but one which was successful. With all the open fretwork there was, one might be forgiven in thinking, a risk that the whole thing could be blown away in any strong wind. However, there was no wind that day, and Frank was as calm as could be, laughing to himself in the anticipation of the reaction from the crowds at the display he was about to give. The engine started smoothly and the two mechanics waved him off, grinning from ear to ear.

Charlie Hands was well attuned to the sound of aircraft and long before his photographer, he recognized the familiar drone approaching from Wapping. He was perfectly placed, in full view of Tower Bridge from London Bridge. No, never, not sensible Mr McClean, surely not? Cecil Grace might have, Mr Charles Rolls might have, but never Mr Frank McClean. As the aircraft rounded the corner from Wapping, Charlie Hands knew what was going to happen and yelled at the photographer:

"FOCUS ON TOWER BRIDGE – NOW!"

People on Tower Bridge turned to statues as the tiny craft approached and, open-mouthed, they ducked as the new seaplane, S-33, flew through the middle of the Bridge, between the roadway and the upper walkway.

Frank waved, and suddenly there was a huge cheer and, to the astonishment and delight of the crowds, with his adrenalin flowing, Frank flew underneath London Bridge, Southwark Bridge, Blackfriars Bridge and, as he rounded the bend in the river, he flew underneath Waterloo Bridge. By now the crowds along the Embankment had been alerted and he was in full view of those on the bank opposite the Houses of Parliament. He finished with a flourish, flying underneath Westminster Bridge, and landed neatly at Lambeth Bridge, right beside the terrace of the Houses of Parliament.

Frank acknowledged the roars of applause all along the Embankment with a triumphant flourish of his white silk scarf and of course a wicked grin. There was no sign of 'Beaumont' and no one knew why he hadn't shown up, so the glory was all Frank's and the public loved it and so did the *Daily Mail*. Once again Charlie Hands had got the scoop of the time. When the dumbfounded Members of Parliament and the public who witnessed Frank's outrageous feat realized that this had been performed by one of their own pioneers, in one of their own seaplanes, attitudes swiftly changed.

Frank McClean lands at Westminster Pier in view of the Houses of Parliament, August 1912 (Courtesy of the Royal Aero Club)

One can imagine the celebrations in Eastchurch that night. Unfortunately Frank had to be towed home, but nothing could dampen the excitement of the result of the achievement.

For months afterwards aviators repeated the feat to a rapturous public until at last a General Seely, the then Minister for the War Office, demanded a ban on such aerial "frivolities". But by then the job had been done, and the public were all too aware of just how valuable an asset the aeroplane was. In due course the various Government departments became more supportive. Training and instruction was now carried out in earnest through Government channels, and Frank and the Shorts no longer had to lend and fund the training. Frank relaxed at last, and was able to follow his own career, sure in the knowledge that Lt, now Commander Samson and Captain Sueter would look after the recruits and carry out and supervise all the necessary testing.

"Come on Horace, old boy, time you had a break from Eastchurch!" Frank announced to Horace one day. "I'm mounting an expedition down the Nile which I would like you to join, and I would like you to build me a special seaplane especially for it. What do you say to that, eh?"

Horace gave it two minutes' thought. "Why not indeed! I've always wanted to see the Pyramids. I'll talk to Eustace and The Kid."

"I've already done that," laughed Frank "and they said they were perfectly capable of holding the fort on their own without you. Probably be glad to see the back of you for a bit, old boy!"

"Oh really!" laughed Horace, but nonetheless felt guilty, and resolved, when he returned from his travels, to work on Charles Rosewarne once again, to get him to move to Rochester with him.

CHAPTER TWENTY-FOUR

Storm clouds approach

In 1913 the Shorts did indeed build a special seaplane (S-80), a four-seater expedition pusher type biplane for Frank McClean for an aerial expedition to the River Nile from Alexandria to Khartoum to take place in 1914. Frank invited Horace, Alex Ogilvie and Dr Norman Lockyer. In retrospect, a curious time to plan and mount an expedition!

As confidence grew in the aviation business, the Shorts' designs and technology progressed steadily. In 1913 their seaplanes were fitted with much more powerful engines, and they also began testing the viability of dropping torpedoes. At about this time Captain Sueter (the live wire at the Admiralty), Oswald recalls,

> conceived the idea of the torpedo-plane. In great secrecy, a seaplane was designed by a technical officer working in the Admiralty under Captain Sueter's direction. The contract to build it was placed with the Sopwith Company. Mr Winston Churchill (then First Lord of the Admiralty) had sanctioned the building of the seaplane, and pressed forward its construction with all his forceful personality.
>
> In its first trials the machine failed to take the torpedo off. Mr Churchill was getting impatient and had specified a date on which the torpedo must be flown and dropped.
>
> Captain Sueter wrote to ask Horace to pay him a visit at the Admiralty, but he sent me. Captain Sueter explained to me the position, and asked me if we could take the torpedo off with our 100 hp Gnome engine seaplane within fourteen days.
>
> Cautiously I replied that we could try. He replied: "If you will say you will try you can do anything." With this encouragement

> I returned to Eastchurch and reported to my brother the conversation I had had with Captain Sueter. Horace at once put in hand two new main float cross bars, bent upwards in the middle, to allow the torpedo to be swung below the main floats clear of the water and fitted with a quick release catch to release the torpedo. With the late Gordon Bell, who was our test pilot, and Mr Bibby, one of our foremen, I went to Calshot Seaplane Base and erected the machine in a few hours . . . The torpedo with a dummy head weighed 700 lbs. By the time our machine was ready to take off it was already dusk. There was much activity, too, with the torpedo-plane which had been specially constructed for the purpose (by Sopwith). It was at the water's edge and mechanics were working on the engine.
>
> At Bell's first attempt to take off, the Short seaplane did not gain enough speed, even to straighten out the ailerons. I heard one of the rival mechanics say, "Safe as a house!" Gordon Bell returned to the shore and reported one cylinder misfiring. This was soon put right. At the next attempt, the seaplane took off, the torpedo was dropped and a magnesium flare attached to it enabled it to be found and picked up.
>
> It was now dark. This was the first torpedo taken up and launched from a seaplane in Great Britain, only a few days before war broke out!

At this time, activity was rife throughout the country. Lord Northcliffe, always a staunch supporter of aviation, campaigned strongly through the *Daily Mail*, supporting the development of seaplanes and advocating that Britain's best defence would be through the deployment of seaplanes in the most suitable strategic parts of the country. To support and encourage his views he organized lucrative prizes for seaplane races, including a £5,000 prize for a 1,540-mile round-Britain seaplane race.

Unfortunately it was not a success, as there were only four entrants: Harry Hawker with a Sopwith, James Radley with a Radley Moorhouse, Frank McClean with a Short aircraft and Samuel Cody with his own biplane design with three floats. Sadly, before the race took place Cody was killed flying at Laffan's Plain, Farnborough. He was buried with full military honours at Aldershot Military Cemetery.

Towards the end of 1913, not only did the Navy expand their Naval Air Station in Eastchurch, but the Army also set up camps there which meant that all the increased activity was hindering the Shorts' heavy workload, and making demands on their productivity. They desperately needed to be near a calm, uninterrupted stretch of waterfront, which they found in Rochester on the Medway. By now Oswald had shown that he was an astute businessman and a good judge of character. His men respected his leadership and his judgement, so it was entirely natural that he should take charge of the works when they moved to Rochester. Eustace still remained in charge of the balloon works in Battersea, but divided his time between Horace and Oswald. Given the combustible nature of the brothers when they were together, it was probably a sensible arrangement.

Frank McClean's expedition to the Nile in 1914 was much enjoyed by everyone who went. Horace was a wonderful companion and a great asset to the small band of explorers, his knowledge of the ancient history and the geological background was extraordinary, and his wonderful sense of humour kept everyone amused. It was a very welcome and much-needed break, albeit not too well timed given Europe's turmoil in 1914. As fond of his fellow pioneers in Eastchurch as he was, he did miss the light-hearted Cecil, the ebullient Brab and the enigmatic Charlie Rolls, who was not amusing or even fun like Cecil and Brab, but a marvellous eccentric. Frank sighed; how things had changed and developed. What would Charles be doing now if he were still alive? What news would greet them on their return? Would Europe be closer to war than ever? Horace was convinced that they would be at war by the end of the year. It was a depressing thought and one which Frank hated to dwell on. He would have to give up astronomy, do his duty, and possibly join up. What a thought. He puffed harder at his pipe.

The ferry began to slow down as it entered the harbour. The engines roared as the brake chains churned, and they had to shout to each other to be heard. They could see large placards on the quay but could not read what they said. The rich experiences of the expedition faded as Horace was brought back to earth by the realization that they were almost home, and the resumption of his work was now urgent. Oswald and Eustace could not be expected to hold the fort for ever. The decision whether or not to go to Rochester still loomed. Horace lifted his hand to shade his eyes so that he might decipher the words on the placards.

"Can you see what they say?" he said to Frank. Frank screwed his eyes up straining to see the words.

"I think it says that someone has been shot!"

They drew closer and the words on the billboards became all too plain.

ARCHDUKE FRANZ FERDINAND AND HIS WIFE ASSASSINATED IN SARAJEVO

ARCHDUKE FRANZ FERDINAND AND HIS WIFE SHOT IN BROAD DAYLIGHT IN SARAJEVO BY BOSNIAN TERRORIST

"This is the beginning of it all," said Horace quietly. He took out his pocket watch and wondered impatiently how long it would take to get back to Eastchurch. Frank looked sideways at Horace and saw how serious he had become.

During Horace's absence more servicemen had arrived in Eastchurch and been billeted around the village. Every householder had a soldier or two staying with them now. When Charles Rosewarne came home at lunchtime Kate was shocked when he told her that Mrs Ross would have to have a cot in the kitchen because they had two more young officers arriving soon.

They faced each other across the dining table. Charles shook his head. Poor Kate, how her world had been turned upside down, her values shaken by events over which she had no control. It surprised him that she still wanted to stay in Eastchurch. What was she trying to cling to?

"Sorry girl, we ain't got no choice!"

He was about to go back to work when the telephone rang, which made Kate jump. She hated the instrument and really could not think why Charles had insisted on having it installed. They had existed all these years without it. Charles picked it up, and as he listened he grinned, laughed, shook his head and suddenly appeared to be concerned, then replaced the receiver.

"That was Horace. He's back. He said Archduke Ferdinand and his wife have been murdered in Sarajevo, and we'll soon be at war."

"Nonsense!" said Kate firmly. "The death of some unknown middle European duke cannot possibly start a war."

Horace's return, at the same time as the news from Europe, and the gathering storm clouds making war seem almost inevitable, had put unbearable pressure on Eustace and Oswald. They hardly had the capacity to fulfil the orders and demands of the Admiralty, but they were thankful for the extra space and facilities which Rochester provided. When Horace joined them matters would improve further with them all being in one place.

"Enough!" exclaimed Eustace, seeing Oswald, overcome with exhaustion, leaning on the drawing table for support. All day long Oswald had had visitors, from the Army, from the Admiralty, from newspaper reporters, from his craftsmen wanting guidance, and so on and so forth, he thought. This was the first opportunity for him to actually complete a job he had wanted to finish two days before. It was nine o'clock at night, still light and no one to interrupt him apart from Eustace. The sooner Horace moved to Rochester the better. He supposed it was the wretched Baynes woman who was keeping him in Sheppey.

He sighed deeply. "You're right, I'm not achieving anything." He put down his pen and put the drawing board away, sliding it underneath the work table. He straightened up and rolled down his sleeves. "Come on, let's go and sit outside, it's a nice evening." They walked along the quay and found a bench to sit on in the late evening sun, overlooking the dock where the two seaplanes, almost completed, were lapped gently by the water.

The two brothers stretched their aching limbs almost simultaneously. Oswald flung his arms wide, bent his elbows and rested his head back on his arms, while stretching out his long legs in front of him. Eustace lit a cigarette and lent back inhaling the smoke, relaxing as he crossed his legs.

"You know, when war is declared, I want to join up!" said Eustace quietly.

Oswald released his head from his hands and sat up sharply. "You will not be able to do that, Eustace, you are too valuable."

"Me? Valuable? Nonsense! You and Horace can manage perfectly well without me."

"That is not true, you are irreplaceable. You have always underrated yourself. It was you who started all this in the first place, and I could not do without you." Oswald searched for the right words. "Eustace, someone, some force, perhaps God, I don't know what drives us, but we have a responsibility now that is greater than our own personal lives. I do not know what this force is, yet I remember all those years ago when I was an errand boy riding the delivery bike on hard, cobbled streets, sometimes the load was so heavy I thought I should fall, but I didn't, despite the load. I had to complete the round no matter how hard it was, fighting against rain, snow, and slippery cobbles. Well, that's how I feel now: I'm driven, I don't know by what. I do know that Short Brothers will be a respected and trusted name, throughout the world, you'll see."

Eustace looked at Oswald. His face was perfectly composed and calm. His dedication to "The Short Brothers" would never falter, no matter where events led. He had become their leader.

It was a combination of factors which had made Horace hesitate to move to Rochester. Edith was one reason, of course. She still had the Britannia to run and it was busier than ever. He could not ask her to give the hotel up, for what else could he offer her at the moment? He was working all hours and saw no let-up on the horizon. Horace wanted Charles Rosewarne to move and build his sheds for him, but Charles could not commit himself. Horace had come to rely on him as a friend. He needed someone he could relax with and be frank with, and know that nothing would go any further. He had never needed this prop before, but with his family life in tatters and the friction between himself and his brothers, he could not contemplate moving to Rochester on his own.

Bert Chesson had told Charlie he would not go, and he told Charles that he was sure none of his tradesmen would want to go either; they were too comfortable in their cottages that Charles had built for them. They were Sheppey men and that was that. Charles Rosewarne had worked with Bert for too long, and the idea that he would have to find another chief joiner if he moved to Rochester, and a complete new workforce, he told Horace, was going to be very difficult, given the huge amount of work he already had in hand for the Admiralty in Eastchurch. Added to which Kate had dug her heels in and told him: "Don't be ridiculous, those things will never take off!"

The only person longing to move to Rochester was Dorothy. She hated Eastchurch now that Leonard, Ruby and Francis were no longer there, and if they moved to Rochester they would be reunited. Also, just to compound her unhappiness, her best friend Florence had left Eastchurch. She hated the place being so overcrowded with servicemen, who whistled at her wherever she went; she felt lonely without her friends.

When Cecil was killed, Amy Cunningham immersed herself in her music and gradually over the months she realized that she could return to her profession as a concert pianist; and furthermore her daughter Florence had won a place in the London Conservatoire of Music to train as a professional singer, so they had moved back to London.

Captain Sueter became the last of the principal unsung heroes, in this extraordinary historic story. He recalls in his book *Airmen or Noahs*:

> After the War broke out, we required Mr Sopwith's efforts and those of his factory to produce high-performance machines, then just beginning to show promise. But Hyde-Thomson and myself were quite determined to succeed with a torpedo machine, so I sent for that fine pioneer seaplane constructor, the late Mr Horace Short. When I explained my requirements to him and the great weight that had to be lifted with a 225 hp Sunbeam engine, in addition to pilot and petrol, Horace Short looked at me with a determined grin and said: "Well, if you particularly wish this done, I will produce a seaplane that will satisfy you"; and he did. Horace Short was almost uncanny in his knowledge of what a machine could do, even in its design stages. I never knew him wrong. At times he made me furious with his late deliveries. But in the end he always turned out something far better in performance than I expected, and it was well worth the extra delay.
>
> He paid great attention to details, worked like hell, seldom left Eastchurch, and burnt much midnight oil that few except myself knew anything about. Undoubtedly Horace Short was a great loss to air development, as by the gifts of the gods he was a creative genius. Creative work appeals to the Short family. They have all done well with balloons, airships (rigid and non-rigid), aeroplanes and seaplanes, and if Oswald Short will only concentrate on seaplane work

> and think of nothing else, we shall have flying yachts and flying ships before he leaves, which I hope will be a very distant date, this planet.
>
> Mr Horace Short's promise was fulfilled and he came up specially to see me at the Admiralty to receive my full forgiveness for past delays in delivery of his machines, and my very warmest thanks for building the seaplanes that enabled our pilots to accomplish this great torpedo feat.

By the outbreak of the First World War, original members of the Aero Club, Charlie Rolls, John Moore-Brabazon, Frank McClean, Cecil Grace, Eustace and Oswald Short were no longer experimenting in Eastchurch. Only Horace Short remained and sadly he died in 1917. Their ghosts remain, and a memorial opposite All Saints Church bears their names in commemoration.

EPILOGUE

THE HONOURABLE CHARLES STEWART ROLLS

> "He was, in the highest degree, a practical scientific worker and his place cannot be adequately filled by anyone else."
>
> (*The Aero*, July 1910)

"Dedication and craftsmanship and the pursuit of excellence." These words are written in today's description of Rolls-Royce, leaders in their field of Civil Aerospace, Defence Aerospace and Energy. Charlie Rolls could have written them himself over one hundred years ago. Henry Royce did not turn his efforts to aviation until 1914, years after Charlie Rolls had tried to persuade him to do so, but when he did he was all too aware of Charles Rolls' prophetic predictions.

THE SHORT BROTHERS

Horace Short remained in Eastchurch during the War, developing his seaplanes and producing aircraft for the Naval Air Services to the highest specifications for Captain Sueter. He worked tirelessly until finally the pressures took their toll and he died of a heart attack in 1917. His friend Charles Rosewarne was by his side and ironically so was his brother Eustace. Sir Charles Parsons had commended him for a knighthood in 1916 but not even a posthumous knighthood was awarded. His genius in those early days of aviation was overlooked and unrecognized by the general public and the Government.

Eustace Short was shy and enigmatic but with an exceptionally strong will. Will enough to stand up to Horace and form his Balloon Company with Oswald despite Horace's lack of support in those early days. During the War he still ran the balloon company in Battersea but oscillated to and fro to Eastchurch and Rochester, liaising over the works in both workshops. He was a true sportsman, a good footballer and a talented boxer. He finally managed to acquire his pilot's licence but not until he was fifty years old. He died at the age of 57 in 1932 of a heart attack while flying a seaplane on the Medway. The architect of the original Shorts company, he also was unrewarded, and his contribution to aviation went unrecognized.

Oswald Short, "the Kid", quietly took full command of the company when Horace died. His quiet confidence in his superb workmanship proved beyond doubt his extraordinary inventive talent in all aspects of aviation. He inspired everyone who came into contact with him. The contribution made by Oswald Short to the technical side of aviation was far greater than is generally realized. Oswald pioneered the use of aluminium alloy for the primary structure of aircraft years before it was actually put into practice, and he also pioneered stressed skin all-metal construction. His quiet courage and conviction carried through those, then, revolutionary ideas. He guided his company through two world wars, a depression and a revolution in design and technology. Oswald Short, a quiet and unassuming man, was never given the tributes he deserved despite the fact that his loyal employees fought for years for the recognition he deserved. Nevertheless his legacy has been immense.

THE OTHER PIONEERS

John Moore-Brabazon – Lord Brabazon of Tara

John Moore-Brabazon loved life to the full and found lovable traits in everyone he came into contact with. History relates that he was wonderful company, always made people feel special and he was definitely a true British hero. He held the No. 1 Pilot's licence from the Aero Club, he adored the Cresta run and was winner of the Curzon Cup three times. He was Captain of St Andrews, the Royal and Ancient Golf Club, and was an accomplished sailor. He served in the Army, achieving great success as a talented aerial photographer, which subsequently led him into commerce with Kodak, and was an effective Member of Parliament for many years.

Frank McClean – Sir Francis McClean

Most generous benefactor, offering sound advice and leadership to the early flying pioneers, always approachable, loved and respected by everyone. In August 1914 he joined the Naval Air Services and was put on Channel patrol and subsequently became chief instructor at Eastchurch until 1919 when he resigned. Joint founder of the Norman Lockyer

Observatory at Sidmouth. He became president of the Royal Aero Club twice and Lord High Sheriff of Oxfordshire in 1932.

Captain Murray E. Sueter – Rear Admiral Sueter, CB, RN

Unafraid of the Sea Lords and their intractable attitudes he was a pioneer of Naval Aviation and oversaw the creation of the Royal Naval Air Services. Despite his obvious ability as a brilliant and respected officer his clash with his superiors found him relieved of his job in 1917 with no work for two years until he was retired as a Rear Admiral. He became MP for Hertford in 1920 until 1945 and was knighted in 1934.

EASTCHURCH

In 1914 only Horace remained of the original and earliest flying pioneers. The influx of servicemen in both World Wars ended and the Army and Naval stations were disbanded for good after the Second World War. The village is now much as it used to be in 1908 apart from the construction of a prison built on the outskirts of the village, which does not impinge on the village itself. One can visit All Saints Church and see the stained glass window dedication to Charlie Rolls and Cecil Grace, "Prisoners of Hope", the Pioneers Memorial opposite the church, Parsonage Farm, Shurland Hall, the Shorts' flying sheds built by Charles Rosewarne, Standford Hill where Charlie Rolls taught himself to fly, the original flying fields bordered by the River Swale, and listen to the marsh birds and the ghosts of the Short Wright biplanes and maybe Frank McClean's Shorts S-33 biplane which he flew through Tower Bridge.

THE CHILDREN

Leonard and Francis Short both attended Rochester Grammar School. Leonard went on to work for Oswald while Francis later returned to Sheppey and took up farming. Both the brothers remained friends with and in contact with Dorothy for the rest of their lives. **Florence (Flo) Cunningham** became a professional singer. She never married, but she and Dorothy remained best friends right up to the days they died. **Dorothy** left Eastchurch in 1917 and was married briefly to a barrister and Welsh Guards officer. It was a very unhappy time but subsequently she remarried very happily and lived in Gillingham, Kent.

ACKNOWLEDGEMENTS

I am grateful for the help derived from the following:

Pioneers of the Skies by Michael Dunne, OBE
Rolls of Rolls-Royce by Lord Montagu of Beaulieu
The Brabazon Story by Lord Brabazon of Tara
Airmen or Noahs by Rear Admiral F. Sueter, CB, RN
Early Birds by Air Commodore Bill Croydon, CBE, RAF (retd)
History of the Isle of Sheppey by Augustus Daly
Royal Aero Club Gazette 1901–1951

I would like to thank Andrew Dawrant of the Royal Aero Club; Norma Crowe, secretary of the Shorts Commemoration Society at The Medway Archives and Local Studies Centre, Rochester; Mo McFarland, the *Daily Mail* library manager; and the Fleet Air Arm Museum, all of whom have provided me with valuable material and photographs.

I would also like to thank my husband John Landau for his patience and encouragement; Simon Shreeve for his valuable contributions regarding the Shorts, his editing notes and his support; Tom Irlam for his support and providing Shorts' memorabilia; Natalie Painter of The Copy Shop, Cheltenham for numerous technical services; and Randall Northam for his advice and assistance and without whom this book would not have been produced.